The Universe Today
ULTIMATE GUIDE TO VIEWING
THE COSMOS

The Universe Today ULTIMATE GUIDE TO VIEWING THE COSMOS

Everything You Need to Know to Become an Amateur Astronomer

DAVID DICKINSON

WITH FRASER CAIN, publisher of Universe Today

FOREWORD BY DR. PAMELA GAY, ASTRONOMY CAST

First published in 2018 by

Page Street Publishing Co.

27 Congress Street, Suite 105

Salem, MA 01970

www.pagestreetpublishing.com

Distributed by Macmillan, sales in Canada by The Canadian Manda Group.

22 21 20 19 18   1 2 3 4 5

ISBN-13: 978-1-62414-544-5

ISBN-10: 1-62414-544-2

Library of Congress Control Number: 2018947012

Cover and book design by Laura Gallant for Page Street Publishing Co.

Photography research by Fraser Cain. All photos/graphics by David Dickinson except those listed on pages 230–233. Star charts created using Stellarium.

Printed and bound in the United States

TO MY MOM,

FOR INSTILLING A SENSE OF THE WONDER OF NATURE IN US KIDS, AND ALL OF THOSE SUMMER NIGHT VIGILS AWAITING THE PERSEID METEORS UNDER DARK NORTHERN MAINE SKIES.

CONTENTS

FOREWORD

It's easy to say, "Hey, everyone should go out and look up." Heck, I say some version of that sentence five to ten times a week! What is hard is finding what to say to get people to keep looking up. It turns out there isn't one key sentence, but there may be one key book—this book—that helps people find constant inspiration to explore our skies.

Astronomy is the rare science that has a story to tell: a story that begins with the formation of our universe, extends into the now of stars and planets and galaxies, and has a predictable end that theorists love to reveal with their scientific spoilers. Astronomy is also one of those sciences that provides amazing images that are as mind-blowing as its scientific insights. This combination of stunning pictures and heart-pounding descriptions of formation and destruction has me captivated for life. Astronomy is the ultimate page-turner, as it describes our own possible fates as people, as a planet, and as a universe.

As a scientist, I've spent my life learning how to read this story with telescopes and space probes, and with each new technology, I and my colleagues find a few more chapters of this epoch-spanning narrative. Today, we are in a new, third renaissance of astronomy. Four hundred years ago, Galileo got us looking up with telescopes during the actual Renaissance, and his efforts and the efforts of all who followed brought us new concepts about celestial motions and the actual nature of planets, comets, and the heavens beyond. In the early 1900s, we entered the age of big glass, glass plate photography, and an expanding universe. Data drove us to accept relativity, and to explore the possibility of dark matter. Today, we've gone digital, and our advances in spacecraft engineering and in multi-mirrored and mega-mirrored telescopes as well as digital detectors have brought us into a third age of discovery, one that is bringing us a wealth of alien worlds, black holes, and more invisible forces and features to shape a space that we find is accelerating itself apart.

While astronomy in the past was a thing to be done by sponsored scientists at universities and by the wealthy who used their leisure time to study, today's astronomy is open to the masses. In fact, it's now possible to take a better image with a simple telescope and an iPhone than I could take as a college student with access to professional equipment!

In 1991, high school me was able to intern at a multi-meter telescope facility. At that cutting-edge facility, we celebrated taking basic digital data and dreamed of upgrading to an 800 x 800 pixel (0.65 megapixel) camera like those on the Hubble Space Telescope. At that point in time, polar aligning wasn't aided by GPS, astrophotography meant using film, and the guider was the observer (or the student like me!) who manually kept things lined up. Today, many clubs and individuals have GPS-driven telescopes, low-end Digital Single Lens Reflex (DSLR) cameras can take images with more than 20 megapixels, and guiding . . . well, a good motor doesn't need any, and if it does, there are digital systems to replace the digits of students.

It is possible for you—yes, you, the person reading this sentence right now—to potentially work in your driveway to study evolving dust storms on Mars, to measure the passage of exoplanets around distant stars, and to take stunning images of far-off galaxies that have your friends asking, "Is that from Hubble?" It's not easy, but you can do it, and this book is here to help.

Today, telescopes are available for almost all incomes, with fun systems starting for less than $100. Unfortunately, many of the telescopes that get purchased go on to live in closets and the corners of garages. This book is here to help telescopes everywhere escape this fate. It does this by packing a one-two punch of science and observational how-to.

On the side of science, *Ultimate Guide to Viewing the Cosmos* brings you stories from professional researchers who work with NASA, ESA, and other spacecraft and telescopes to revolutionize our understanding of the sky. The author will tell you about what they are discovering and how. These interviews are illustrated with never-before-seen images from some of the world's best astrophotographers. These are the kind of picture-stories that will make you want to both learn more about astronomy and look up more to see what you can capture for yourself.

To help you sate your observational hunger, this book lays out all the information you need on when, where, and how to observe the sky. From tricks for enjoying a meteor shower or solar eclipse to advanced details on celestial motions and color imagery, it offers something new to try for people of every skill level. This is the kind of book you can read and then recommend to a friend or family member just starting to get interested in astronomy, or that you can pass to a longtime amateur astronomer looking for inspiration on what to try next. Heck, this book has things this professional astronomer wants to try with her personal scope!

So, go out and look up, and let this book inspire you to keep looking up and to unlock your own chapter of the mystery of astronomy.

—Dr. Pamela Gay, Astronomy Cast

INTRODUCTION

Stargazing is one of the oldest forms of entertainment there is. For millions of years, humans looked at the stars, carefully noting the motions of the Sun, Moon, and planets as they made up stories about the patterns in the sky. In a way, we've lost much of the intimate everyday knowledge our ancestors had about the sky. In the modern age of distraction and flashing electronic lights all vying for our attention, the act of gazing skyward is a minor stroke of rebellion.

What do you see, looking at the night sky? How would you make sense of it all, if you were simply plopped down on planet Earth?

We don't often take the time in our modern, fast-paced society to simply stop and observe the world around us. We should. There are strange and wonderful things in your own backyard, if you know where to look for them. You can do this simple act of discovery from a country farm or a city rooftop, no telescope or camera required. You're approaching the sky as our ancient ancestors did, without any preconceived notions.

Of course, we're lucky to live in such a privileged time. We now know that stars are suns, but far away. We know that our sun is a vast fusion furnace, about halfway through its 10-billion-year main sequence life. We also know something about our Milky Way Galaxy's address in the Orion Spur, here in our home Solar System on planet Earth.

We all share the same vantage point in time and space as we hurtle around the Sun, orbiting in our Milky Way Galaxy in an ever-expanding universe. We're also privileged to stand on the shoulders of giants, generations of sky-gazers who have done the hard work of mapping out our little patch of space and time. Thanks to them, we know something about the cosmic address of planet Earth, as we enjoy the **stelliferous era**—the current cosmological age of the Universe when stars can shine and create heavier elements, including the carbon, nitrogen, and oxygen necessary for life on Earth—13.8 billion years after the Big Bang.

But how would you guess at all of this, looking at the sky tonight? One by one, as the twilight sky gets darker, you can see the stars emerge. Perhaps watching from one evening to another, you note how some of those stars "wander" against the starry background, as planets in our Solar System betray their presence. After a while, a meteor might silently punctuate the night. Perhaps Earth's moon will rise, a cosmic companion that has opened up so many of the Universe's riddles to us.

Watching the sky over a span of weeks and months, seasons will change as the sky appears to shift and the length of day versus night changes. Watch long enough, and you just might see a hairy star, a comet visiting the inner Solar System from the depths of the dark outer Solar System and the Oort Cloud.

Closer in, you might see other "stars" drift by, as the contemporary age of satellites make their modern mark on the ancient sky.

Welcome to the wonderful world of astronomy, a lifelong interest where there's always literally something new under the Sun to learn. Perhaps you'll just be happy knowing what you are seeing with a casual glance at the night sky, or maybe you'll get bitten by the astronomy bug and want to pursue the interest further with a telescope or a Digital Single Lens Reflex (DSLR) camera. Hey, you might even want to do real science, and add to that body of human knowledge using observations done in your very own backyard.

That's the intent of this book: a guide to navigate the path through amateur astronomy from entry-level beginner to expert. Back in the day, astronomy and science were the realm of the rich elite, as backyard tinkerers with the time and patience pushed the bounds of discovery. William Herschel, for example, discovered the planet Uranus from his backyard garden in Bath, England. Galileo's first crude telescope represented the very cutting edge of early seventeenth-century astronomical technology, and is easily surpassed in optical performance by a toy telescope today.

Now, more than ever, the world's information is at the fingertips of a budding amateur astronomer. Prices have also sharply dropped for quality telescopes, and amazing sky photos are as easy as plopping a DSLR camera on a tripod and doing timed exposures of the night sky.

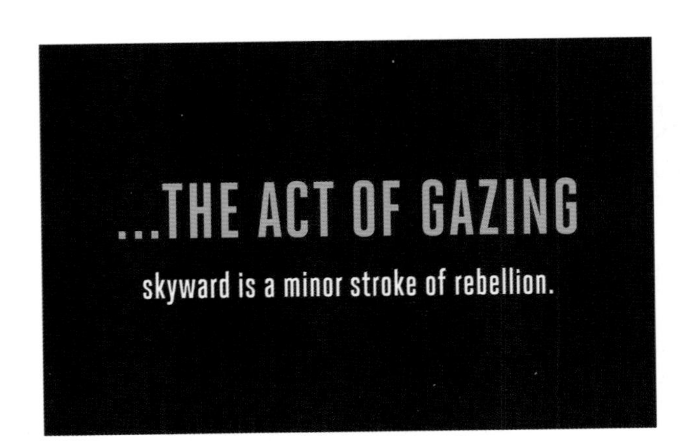

Conversely, the bar has been raised in astrophotography, as we now routinely delete photos that would've been award-winning a decade ago. It's easy to get discouraged and forget to simply enjoy the night sky, instead of thinking you need a $10,000 camera and telescope rig to compete with the pros.

True, some people like to maintain a certain mystery to the Universe. We feel that knowing, say, that a faint speck of light in the eyepiece is actually a quasar of stupendous energy seen across the Universe from billions of years ago—as well as how we know all of this, in a coherent view of the modern Universe—adds to the wonder of reality, rather than detracts from it.

In this guide, we'll trace out not only our place in time and space, but "how we know what we know." We'll look at our evolving understanding of our place in the cosmos, and how that's changing even today.

We will also give you the eyes and knowledge to interpret the sky, as we look at the seasons and the planets month-to-month and move out to deep sky observing. We'll look at binoculars, telescopes, and cameras for deep sky astrophotography, and how to navigate the problems and pitfalls of purchasing that first telescope that's right for you. Telescopes are like tools, and there's no perfect one-size-fits-all for the job.

We'll also look at eclipses, auroras, satellites, and some of the stranger phenomena you'll encounter in the night sky. And we'll let you in on a little secret: amateur astronomers never see UFOs . . . this is because they're intimately familiar with the night sky, a skill you too can acquire. They're out every night, always on the lookout for a surprise meteor or auroral outburst. Some events in the sky, such as eclipses, conjunctions, and occultations, happen like clockwork and are known years in advance. Others, such as comets or meteor showers, can surprise us, adding a bit of celestial drama to routine sky watching.

And speaking of UFOs, we'll also look at the science of astronomy and knowing the night sky as a discipline for critical, evidence-based thinking in an increasingly anti-science world. Astronomy, cosmology, and climate science all have their own pseudoscience twins, from astrology to moon-landing deniers to flat-Earth conspiracy theorists. Next time a coworker shares a fake astro-photo that's *too* perfect to be real, you'll be able to spot it immediately with an expert eye.

Astronomy doesn't end at sunrise. We'll not only delve into how to safely observe the Sun, but also how to observe the Moon and even planets in the daytime. Many folks are amazed when I show them Venus through the telescope on a busy urban sidewalk, unaware that such a feat of visual athletics is even possible. I've even had folks stubbornly refuse to believe that a tiny speck of light such as Venus is actually another *world*, the size of the Earth!

Clouded out? The Web has also become a great resource, and lots of astronomy has moved online. You can access telescopes online, and conduct citizen science projects to hunt for exoplanets, supernovae, and more. Many guides to astronomy leave off where the Web and smartphone technology begins, which is unfortunate. We'll look at resources, projects, and the fast-changing world of astronomy online.

And speaking of projects, there are lots of DIY projects that can be done on the cheap, from solar projectors to cardboard interferometers. You can, for example, start imaging planets tonight, using nothing more than a laptop, a modified webcam, free software, and a modest telescope. Want to build a telescope for yourself? In the olden days, this was the only way to gain access to a large telescope. Even today, building a telescope not only allows you to gain a deeper understanding of how optics work, but also allows you to incorporate desired design features that aren't available in mass-produced instruments.

You may even want to take the next logical step and build an observatory to house that telescope. This will allow you quick access—while spending less time lugging equipment to a remote dark sky site and setting up gear—and more time actually observing.

Or maybe you simply feel that your backyard is your very own open sky observatory. Whether we're on the beach, a hilltop, or an urban sidewalk, there's always something to look for in the sky, if you know when and where to look for it.

Unfortunately, less and less dark skies are accessible to the general public. Few have seen the Milky Way for themselves, and many fail to realize what most of us have lost in the short span of a generation. Light pollution doesn't just impact the beauty of the night sky, but it's also an economic and health issue as well. We'll look at the problem and provide tips on not only how you can make your property dark sky–friendly, but how you can convince your neighbors and community to adopt the same practices.

And speaking of community activism, one of the very best ways to get involved is to attend or volunteer at a local star party. This is a great way to look through telescopes before you buy one, and meet other amateur astronomers in your own community. I also find that public outreach helps to keep my mind sharp, as I explain everything from the latest discoveries or how far away that distant galaxy really is.

Finally, we'll explore when to look. I often get questions like, "How do you know that 'moving star' is the International Space Station?" or "How do you know where to find the Andromeda Galaxy?" The answer is one of the most fundamental parts of observational astronomy, so basic that many longtime observers take it for granted: the crucial element of time, and knowing not only where and in which direction, but when to look. To this end, we'll provide a six-year look ahead at sky events, including eclipses, occultations, conjunctions, meteor showers, and more. We plan on updating this list periodically, giving this guide an added bonus as a handy and practical reference.

Of course, much of astronomy is about happy surprises, and an event such as the sudden appearance of a new, bright comet can add exciting drama that will sweep across the news. In this guide, we'll let you in on the Web pages and sources we're looking at for fast-breaking astronomy news. Though there's lots to sort out on the web, the information is there if you know where to look for it. We'll show you where to go for the *real* scoop on all things astronomical, and how to make sense of the deluge of information.

Astronomy can be done anywhere you can see the sky. We've stargazed from the darkest skies of northern Maine to the bright streets of the Las Vegas strip, arguably one of the most light-polluted skies on Earth. Though not a prerequisite, astronomy is often enhanced by travel. Hunting for truly dark skies and chasing the path of totality during a total solar eclipse makes travel essential. We'll cover how to find those hidden dark sky gems like a pro, and how to pack lightly for the next total eclipse expedition.

Astronomy is a lifelong pursuit. We always like to say we learn something new about the Universe every day, and we try to challenge ourselves to hunt down at least one new object per nightly observing session.

Perhaps you'll want to venture on to astrophotography and capture images of the night sky for yourself. We'll explore gear, techniques, and how to get the perfect shot.

Want something more?

Astronomy is one of the very few fields where you can still make a meaningful scientific discovery, right from your backyard. With skill, patience, and a little experience, you too can:

- Track satellites (page 116)
- Discover a comet (page 124)
- Monitor meteor showers (page 130)
- Discover an asteroid (page 137)
- Follow solar activity (page 143)
- Track variable stars (page 200)
- Time lunar and asteroid occultations (page 201)

It's a big Universe out there, and there are only so many professionals and observatories to monitor it. Amateur astronomers provide a valuable service, keeping watch where professionals often don't.

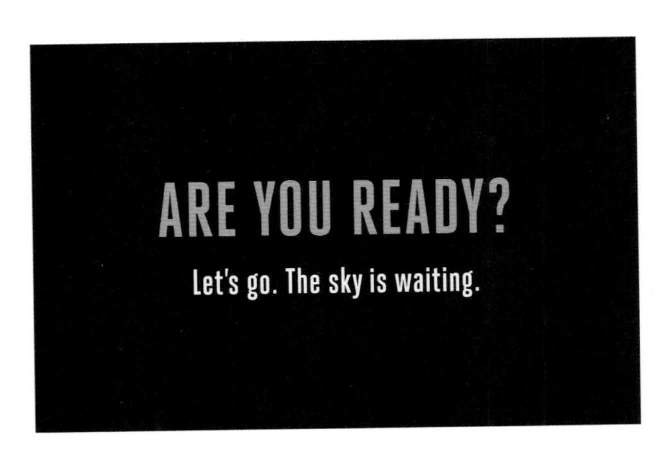

STEPPING OUT TONIGHT

Finding your way in the dark and knowing how to speak astronomer.

"I have loved the stars too fondly to be fearful of the night."

—Sarah Williams

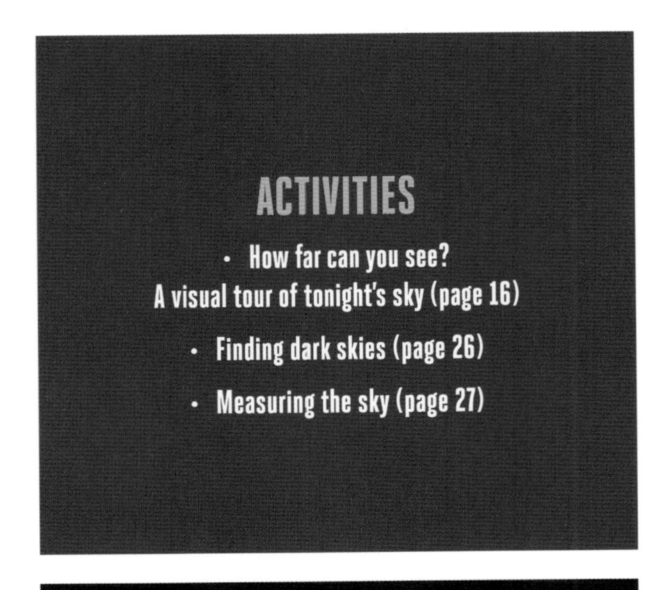

PROTECT YOUR NIGHT VISION.

You're going to need it.

Before you read any further into this guide, here's a challenge: Right around sunset tonight (assuming the skies are reasonably clear), simply step outside and check out the night sky.

What would you see, gazing into the sky?

Once the Sun sets, you enter the period known as **dusk**, a twilight realm where the lighting changes almost moment to moment. Photographers are familiar with this special time. Day shifts to night as the Golden Hour gives way to the Blue Hour. Local objects are drained of their color, as darkness deepens and the cone cells of our eyes switch over to their more sensitive rod cells in the dwindling light,[1] like switching from fast to slow film in daytime versus nighttime shooting. Using an eye patch or sitting in a darkened room can even allow you to dark adapt in advance for an upcoming session of nighttime observing. The better your eyes are dark adapted, the more you'll see. The bright screens of modern electronics can ruin night vision, and many amateur astronomers will cover them with either a red-light filter to preserve night vision, or set apps to adjust the light to a dimmer, reddened night-vision mode while observing. Several apps are even available to automatically dim screens and filter out blue light during twilight hours.

Protect your night vision. You're going to need it.

What do you see tonight? Is the Moon visible? The twilight period during dawn or dusk is actually divided off into three types: **civil**, **nautical**, and the darkest of all, **astronomical twilight**. Perhaps, as twilight darkens, you might spy a star moving slowly across the sky. This is a satellite, a product

The human eye under normal daylight conditions (left) versus 10 minutes of dark adaptation (right).

of the modern Space Age. Satellites, for the most part, move across the sky with the rotation of the Earth from west to east, unwavering moving "stars" illuminated by the Sun in the twilight that soon wink out upon hitting the Earth's shadow. Most of these are old rocket boosters, silently tumbling end over end in orbit. Every week or so, you might just spy the International Space Station. Humans have continued to occupy this orbital outpost in space since the arrival of Expedition 1 in 2000.[2]

Watch the sky long enough, and even more curious sights can appear.

Living at middle and high latitudes, you might just see the flicker of **auroras**, the interaction of the Earth's upper atmosphere with the solar wind, proving that we do indeed live inside the atmosphere of our host star, the Sun.

A **meteor** may punctuate the night, a flashing emissary of a **comet** whose path intersected that of the Earth long ago. These pick up in tempo during annual **meteor showers**, and very occasionally, put on one of the most stupendous displays in astronomy during **great meteor storms**.

Then, one by one, the stars appear. They're there in the daytime as well, when they are overpowered by the brilliance of the Sun. These same patterns visible tonight remain nearly unchanged through historical times, a nighttime backdrop for human drama on which we pin our hopes and fears. The stars flicker as their light—traversing the vacuum of space for centuries—is distorted in its final fraction of a second journey through our atmosphere.

HERE'S A HANDY GENERAL RULE:

Stars flicker or twinkle; planets don't.

You might also notice a few beacons of light, unwavering in their constancy. These are **planets**, whose tiny disks are still just large enough to remain less affected by the roiling convection cells in our Earth's atmosphere.

Here's a handy general rule: Stars flicker or twinkle; planets don't.

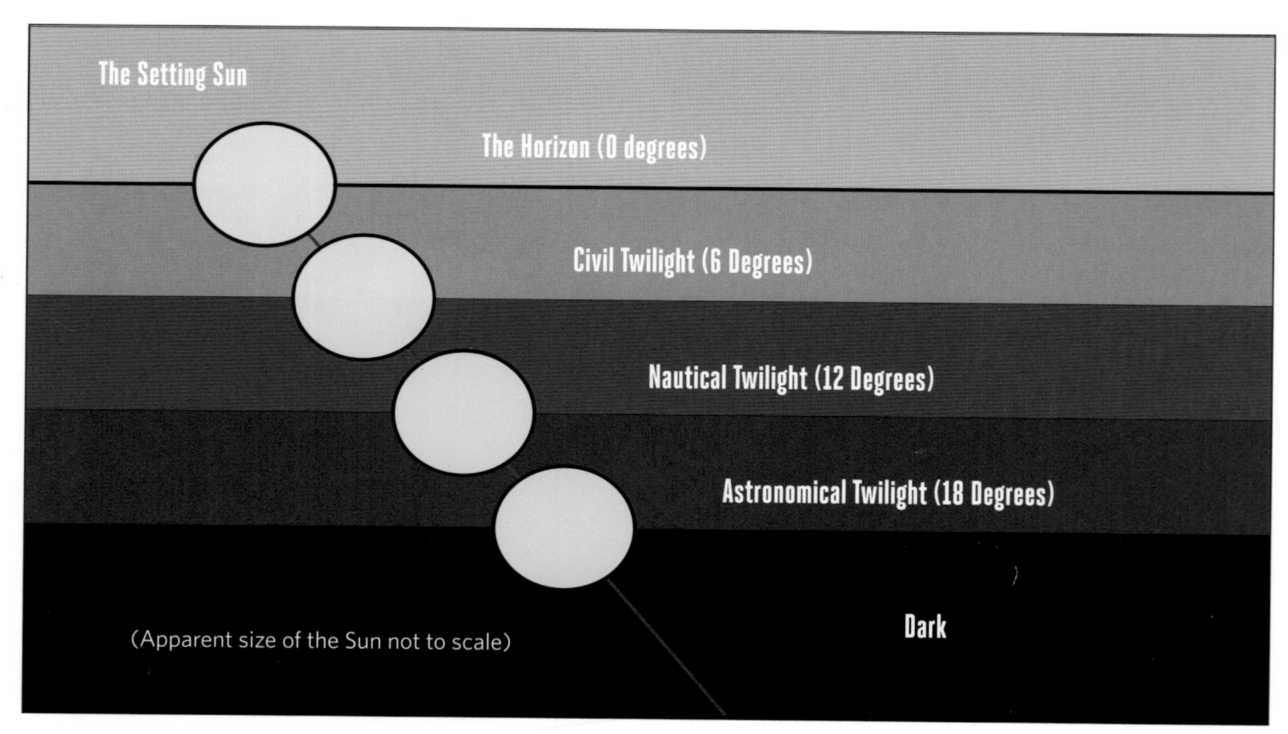

The progression of twilight, from sunset to true astronomical darkness.

MEASURING THE SKY

HOW CAN WE DESCRIBE THE POSITIONS OF OBJECTS IN THE SKY?

How can we describe the positions of objects in the sky? The illusion of the sky as an upside-down bowl overhead covering our little patch of land out to the local horizon is a handy one to understand the motion and position of things overhead. Celestial cartographers used the analogy of a geocentric Earth inside a crystalline, transparent sphere covering the vault of the sky, and even constructed models of this early structure of the Universe as seen from the outside vantage point in space, looking in.

Today, we know this is a fantasy and an illusion, owing to the enormity of the Earth and the Solar System versus the tiny patch of our local vantage point out to the local horizon a few miles away.[3] For example, the curve of the Earth is immediately apparent on the face of the Moon during a lunar eclipse. This curve can be seen at any angle, whether the eclipsed Moon is low to the horizon or high overhead.

Think of the **true horizon** as an imaginary line where the Earth meets the sky, running a complete 360-degree circle around tonight's observing patch. Looking out along the ocean from the beach is the closest to the true horizon that you can ever see, a line straighter than the hand can draw. Of course, this line is indeed curved a *tiny* amount, slighter than the eye can discern. Earth is just that big.

The point directly overhead is the **zenith**, and the imaginary opposite point at your feet is the **nadir**. On Earth, the nadir point is forever invisible, though in space, you can see objects both at the zenith and nadir.

We use **degrees of arc** to measure the distance between one object and another in the sky. This separation is apparent—as in one object may be much closer or farther away than the other—and biased as seen from our Earthly vantage point. The circumference of the sky can be further divided up into 360 degrees of arc, with each degree divided into 60 **arcminutes** and each arcminute into 60 **arcseconds**.

The zenith, for example, is located 90 degrees above the local horizon, straight overhead; the Full Moon is about half a degree (30 arcminutes) across, meaning you could actually stack 180 Full Moons from the horizon to the zenith.

You can make rough measurements of objects in the sky simply by placing your open hand at arm's length. (See this chapter's activity, Measuring the Sky, page 27.) The span of your hand is about 10 degrees across, and each finger is 2 degrees across. Try it next Full Moon: you can actually hide four Full Moon widths behind your thumbnail at arm's length!

Smaller measurements such as arcseconds only come into play when we're talking about views under magnification, such as the separation of close double stars or the apparent diameters of planets. One arcsecond is 1/3600th of a degree, a smaller angle than the human eye can discern. The smallest angle in the sky an eagle-eyed observer might split is just under an arcminute. Likewise, double stars under an arcsecond apart represent a difficult split at the eyepiece under magnification. Professionals further divide arcseconds into milli- (thousandths) and micro- (millionths) of an arcsecond; truly tiny angles, indeed. These minuscule angles only come into play when astronomers measure **parallax**—the tiny shift in a star's apparent position as a function of its true distance—and occasionally, the angular diameters of the very largest stars themselves.

Exploring the upside-down, bowl-shaped illusion of the sky overhead, we can then describe the position of an object using a two-coordinate system, with **azimuth** as the left-right position on the horizon starting at zero (north), to 90 degrees (east), 180 degrees (south), and 270 degrees (west), and the **altitude** or **elevation** as the object's position above the horizon from zero (on the horizon) to 90 degrees (at the zenith).

ACTIVITY: HOW FAR CAN YOU SEE? A VISUAL TOUR OF TONIGHT'S SKY

As mentioned previously, the night sky looks like someone simply overturned a bowl above our observing site. Of course, this is an illusion, and things overhead in the sky are (relatively) near or far (very far) away.

But how far can you see? It's a question often asked at star parties. Let's look at the night sky, with distance in mind.

Space begins about 62 miles (100 km) overhead, above what's known as the **Kármán line**.[4] Those satellites you see zipping by in the twilight sky are actually only about 250 miles (400 km) or so away in Low Earth orbit when they're directly overhead, a short day's drive if you could drive your car straight up. The ring of geosynchronous satellites orbiting the Earth once every 24 hours is farther still, at 22,236 miles (35,785 km) above the surface of the Earth. But that's only about 1/10th of the way to the Moon, almost a quarter of a million miles distant.

Moving out into the Solar System, it's handy to simply drop all of those zeros from miles and kilometers and state distances in **astronomical units** (AU) with 1 AU equal to the Earth-Sun distance of 93,000,000 miles (149,700,000 km). Light, traveling at 186,282 miles per second (299,792 km/s), takes 8 minutes to traverse space from the Sun to the Earth, and just over another second to reflect off of the Moon back onto the Earth.

Distances to the planets take us farther out still. If Mars is out tonight and at closest approach near opposition, it's an average of 49 million miles (79 million km) or 0.53 AU (4.4 light-minutes) distant.[5] Saturn, the outermost naked eye, classical planet, is 930 million miles (1.5 billion km) or 10 AU (83 light-minutes) away.

And Voyager 1—the most distant object hurled into space by human rockets—was, as of October 10, 2017, 140 AU (13 billion miles [21 billion km] or just under 20 light-hours) distant.[6]

But things scale up *again* over a thousandfold looking out towards the stars. All of the stars visible tonight are part of the Milky Way Galaxy, which contains perhaps an estimated 400 billion stars. The very closest star system is Alpha Centauri, about 4.4 light-years away. Up north, you can spy Sirius, the brightest star in the sky at about 8.6 light-years away. Most of the stars visible in the night sky, however, are a few hundred light-years away and much larger and brighter than our Sun.

And things scale up *another* thousandfold, looking outside our galaxy. If you're fortunate enough to live in the southern hemisphere (the South seems to have all of the grand sky objects), then you can spy the cloudy wisps of the **Large** and **Small Magellanic Clouds**, irregular satellite galaxies of our own home Milky Way Galaxy, located 158,000 and 199,000 light-years from Earth, respectively.

But don't despair if you live up north. In fact, you can see a bit farther out into the Universe tonight using nothing but the naked eye by following these simple steps:

- Look past the upper-left corner of the Square of Pegasus asterism and into the constellation Andromeda. (Fall season in the evening is best.)
- Look for the fuzzy patch of the dark sky, 2.5 million light-years away.
- Marvel at how that light left that galaxy back in the early Pleistocene epoch, during the heyday of our ancestor *Homo habilis*.

This represents one of the most distant naked eye objects in the sky.

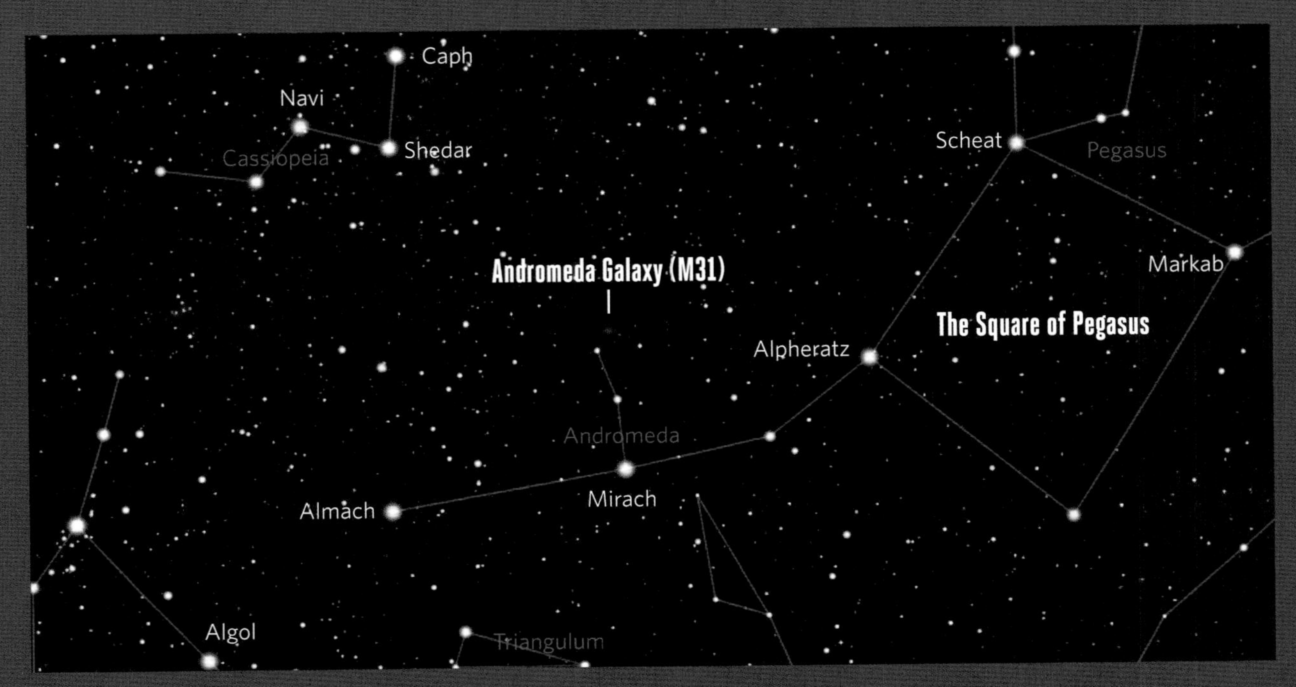

How to find the Andromeda Galaxy (Messier 31) versus the Andromeda constellation and the Square of Pegasus asterism.

THE ASTRONOMER'S NEMESES:

CLOUD COVER, ATMOSPHERE, AND AIR MASS

It's a deceptively simple term. **Seeing**—the turbulence or steadiness of the roiling atmosphere of the Earth above your observing site—will, more than anything else, dictate your observing plans, making you decide whether to set up for a night's worth of observing or pack it in for the night. We're fortunate, in a way: Other worlds, such as Venus or Titan, are permanently shrouded in clouds, with no view of the cosmos beyond. What would an observer evolving on such a world make of the Universe beyond these cloud tops?

Will it be clear tonight where you live? Most modern weather forecasts only paint a partial picture when it comes to local cloud cover. Humidity, temperature, seeing, and **transparency**—a measure of contrast and the ability of fainter objects to punch through the sky glow haze—all play a role.

Amateur astronomers usually assign seeing and transparency a numerical value of 1 to 10, with 1 being a turbulent, murky white sky where stars twinkle and planetary views dance at the eyepiece like you're looking at them from the bottom of a swimming pool, to 10, which represents legendary, rock solid, high contrast, inky black skies with tack-sharp views.

You might see an observer's log reading something like "seeing/transparency = 5/8." These assessments can be a bit subjective, and there are other scales used to assess sky conditions.

WILL IT BE CLEAR TONIGHT WHERE YOU LIVE?

Keep in mind that you're also looking through a thicker section of atmosphere low to the horizon versus the zenith. This is referred to as air mass, with the clearest, steadiest views being straight overhead near the zenith, versus murky, trembling views low to the horizon. Buildings, parking lots, and bodies of water can also add to the unsteadiness, as they re-radiate the heat of the day back out to the sky and space above. Of course, while we might curse the cloudy skies tonight, we can be thankful our homeworld is insulated with a life-giving atmosphere—it would be a very chilly place, otherwise—but the presence of clouds is the bane of many an astronomer. This is why major world-class observatories are located on windswept mountaintops in Hawaii or in the deserts of Chile, places with unparalleled seeing and clear, transparent skies more than 300 days out of the year.

Two great sites to watch for cloud cover forecasts tailored for amateur astronomers are Skippy Sky and the Clear Sky Chart.

THE BORTLE SKY SCALE

In addition to the quality of the atmosphere, light pollution is the bane of astronomers everywhere. We can describe how dark or bright the sky is using what's known as the Bortle Sky Scale,[7] an estimate that you can use for your very own observing site (see section in Chapter 11: How Dark Is My Sky? How to Gauge the Quality of the Night Sky, page 208).

THE BORTLE SKY SCALE		
Number	**Type of Sky**	**Visibilty**
9	Bright urban sky	Only the Moon, bright planets, and bright stars are visible.
8	Urban sky	Familiar constellations may be missing fainter stars.
7	Suburban transition	Night sky is only dark directly overhead, with light domes around the horizon.
6	Bright suburban sky	Milky Way is faintly visible near the zenith.
5	Suburban sky	Milky Way is faintly visible down toward the horizon—zodiacal light may just be seen.
4	Rural transition	Milky Way and zodiacal light easily seen.
3	Rural sky	Zodiacal light extends to the horizon—deep sky objects such as the Andromeda Galaxy easily seen with the naked eye.
2	Dark sky	Limiting naked eye magnitude +6, clouds appear like dark shadows in the sky.
1	Dark sky of an almost mythical quality	Airglow and perhaps the hint of the elusive counterglow known as the gegenschein is visible at the anti-sunward point.

The transition from urban, to suburban, to rural, to a remote site with truly dark skies can be a delicate balance . . . even rural skies, such as this splendid Milky Way scene taken from Dinosaur Provincial Park in Alberta, Canada, shows the light pollution dome of a distant city.

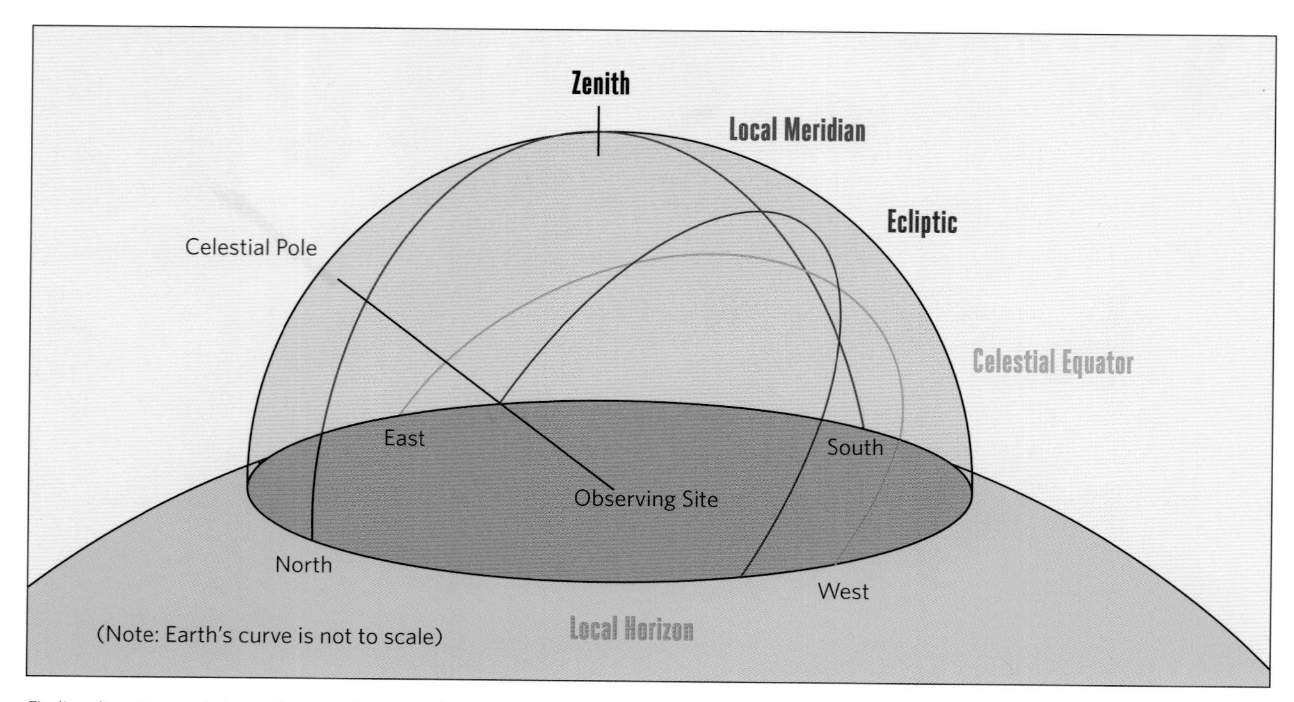

Finding directions and pinpointing coordinates in the sky, using the analogy of the sky as an imaginary "bowl" turned upside down over the observer's viewing site.

SPLITTING THE SKY

To make sense of the sky, astronomers section it off into groups of stars known as **constellations**. Constellations are stick figures representing what we see in space and time from our Earthbound vantage point. The constellations in the sky vary from one culture to another: One culture saw a cart and wagon in the Big Dipper, for example, while another saw a plow. The International Astronomical Union currently uses eighty-eight constellations to segment off the vault of the sky in the northern and southern hemispheres. Some of these, such as the twelve zodiacal constellations girdling the ecliptic date back to antiquity, while the southern constellations are newer constructs dating from the age of exploration,[8] starting with the Bayer family of constellations introduced by the German celestial cartographer Johann Bayer in 1603.

Informal groupings of stars are known as **asterisms**. The Big Dipper, the Summer Triangle, and the Keystone of Hercules are three examples of asterisms. Unlike constellations, there isn't a formal listing of defined asterisms in the night sky; feel free to make up your own.

As with political boundaries, the borders of constellations are human constructs, imaginary lines that help us describe and partition the Universe and the sky overhead. We can say, for example, that a comet is "crossing the border of the constellation Lyra into Hercules" and convey some idea of its position in the sky.

For a more precise position, astronomers use an **equatorial** coordinate system based on the imaginary positions that the rotational poles of the Earth are pointing at in the sky. This system is like the coordinates of latitude and longitude here on Earth. Again, it's helpful to think of the globe of the Earth embedded in an imaginary sphere of the sky.

Declination is the north-south position of a given object measured in degrees north (+) or south (-) above the celestial equator. The celestial equator is declination 0 degrees, while the poles are at 90 degrees north or south. Standing on the Earth's equator, the celestial equator passes directly overhead through the zenith, while at either pole, the celestial equator traces out the true local horizon.

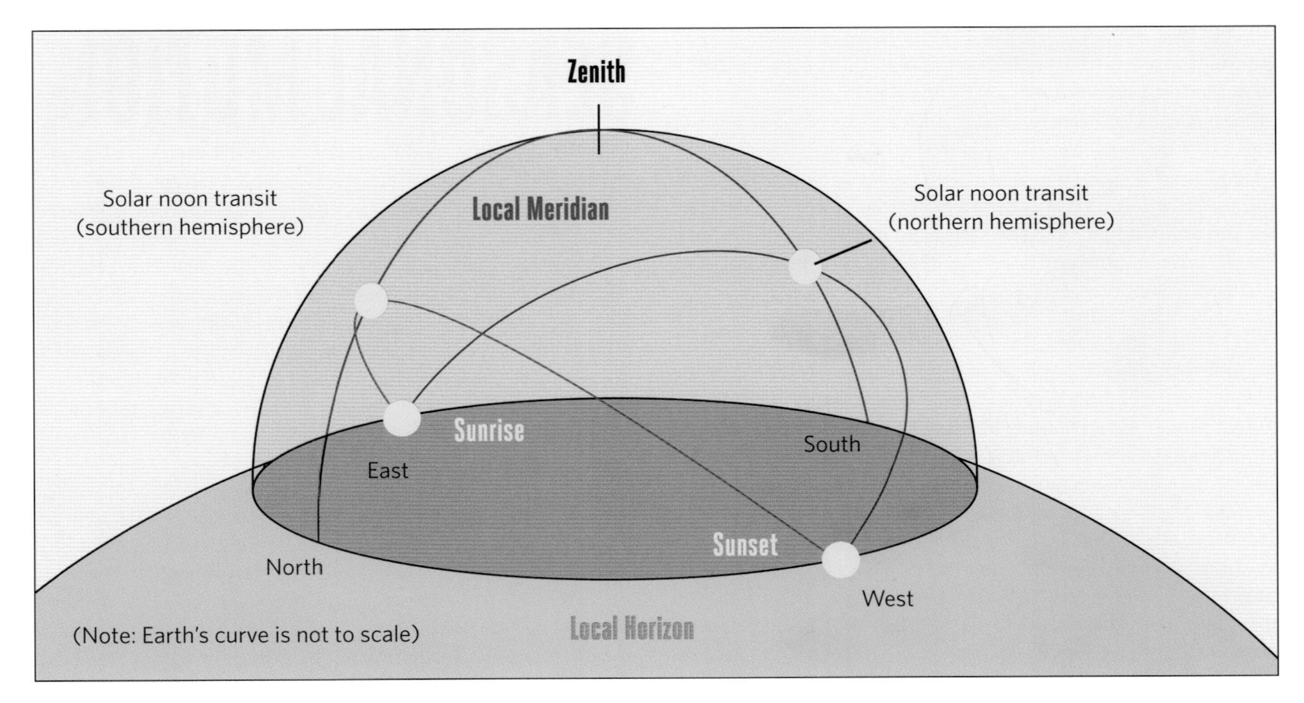

The Sun crossing the local meridian at solar noon; the view from up north versus down south of the equator.

The right-angle analog to terrestrial longitude is **right ascension**, usually simply abbreviated as R.A. The celestial equator is divided off into 24 hours of right ascension, each of which are broken down into 60 minutes and 60 seconds. This system arose as a way to make sense of position (mainly at sea) versus a known rising or setting time of an object from a fixed observatory on land, though it really took the advent of reasonably precise chronometers in the eighteenth century[9] to make it work well.

One note about arcminutes and arcseconds in R.A.: like lines of longitude on the globe, they get narrower the farther you are from the celestial equator. That is, an hour of right ascension along the celestial equator equals 15 degrees of arc, but each arcminute of R.A. equals 15 arcminutes (15/60=0.25 of a degree) of standard arc. Like lines of longitude on the globe of the Earth, these all slim down near the poles. This may be confusing, but these all relate with keeping terrestrial time on your wristwatch or mobile phone in step with observed astronomical time.

The equatorial plane is also handy for Earthbound telescopes. Tip a telescope and point one rotational axis at one of the poles, and it can slowly track the sky overhead using a clock drive, without the need to continually nudge the object back into the field of view.

Finally, in addition to the ecliptic, equatorial, and alt-azimuth reference planes, there are two other systems to consider.

Draw a line through a celestial pole, across the local zenith and the pole's opposite cardinal point (for example, the north celestial pole to due south, or vice versa in the southern hemisphere), and you come up with your **local meridian**. The moment an object crosses this imaginary line in the sky is also known as a **transit**. In the northern hemisphere, for example, the Sun rises in the east, transits due south, and then sets in the west. It's a strange sight for those accustomed to motion up north, however, to journey to the southern hemisphere and see the Sun transit to the north!

Objects that remain above the local horizon wheeling around the poles are referred to as **circumpolar**. At the poles, every object in the sky overhead is circumpolar; at the equator, the reverse is true, with every object rising and setting once a day. Most of humanity lives between these two extremes, and instead sees a grouping of stars eternally circling the poles. The points where circumpolar objects cross the local meridian are known as upper (between the pole and the zenith or due south) culmination and lower (between the pole and due north) culmination.

The second reference system is the **galactic plane**. This imaginary line bisects the plane of our home, the Milky Way Galaxy. We're just over 26,000 light-years[10] from the galactic center, which lies obscured by dust towards the direction of the constellation Sagittarius. Look toward the spout of the teapot asterism, and you're looking toward the mouth of a monster, the supermassive black hole dubbed Sagittarius A* (pronounced "A-star") which is four million[11] times as massive as our puny Sun. We orbit the core of our galaxy once every quarter of a billion years, bobbing up and down through the galactic plane. There's a reason to think that in addition to habitable zones in a solar system—a Goldilocks Zone, where it's not too hot for liquid water, but not too cold—galaxies may have a similar zone suitable for life, where it's not too radiation-riddled (like in the core) but not too metal-poor[12] for elements like oxygen and carbon (the outer rim).

However, the galactic coordinate system is rarely used in Earthbound observational astronomy. Perhaps, if you're a starship captain, it would prove handy (science fiction authors take note), navigating the galaxy using pulsars as cosmic lighthouses.

Got all of that? These coordinate systems all help make sense of describing where things are in the sky. Again, it's important to remember that these systems are all imaginary, a product of our Earthbound vantage point and our way of making sense and describing the location of things in the sky.

THE STARS THEMSELVES ARE ALSO IN MOTION.

SEASONAL MOTION

Watch the sky from night to night and you'll note key changes, just like our ancestors did. One major change is the view of the stars with respect to the Sun from month to month. Follow the course of the Sun through one calendar year, and it returns to nearly the same spot in the sky. The celestial path the Sun follows is known as the **ecliptic plane**, which is actually the trace of the orbit of the Earth around the Sun. The traditional twelve segments of the ecliptic are the signs of the **zodiac**, owing to the science of astronomy's origins in the pseudoscience of astrology, an early attempt by human civilization to ascribe meaning to the motion of the heavens versus terrestrial affairs.

The nighttime perspective of the Earth slowly changes as the planet goes around the Sun, changing the view of the sky beyond. An observer's latitude also determines how much of the total sky wheels overhead: At the poles, an observer sees only one-half or hemisphere of the sky, while at the equator, an observer would theoretically see the entire sky from the northern to southern rotational pole. If you can measure the position of the rotational pole above the local horizon, you can determine your latitude on the Earth.

LONG-TERM MOTION

This view is also slowly changing over time. In addition to our planet's daily rotation on its axis and its yearly motion around the Sun, the axis of the Earth is also slowly wobbling, taking about 26,000 years to complete one wobble. This means the celestial poles trace out a 47-degree-wide circle in the sky in what's known as the **precession of the equinoxes**.

Though the bright star Polaris is currently pole star for observers in the northern hemisphere, the star Thuban in the constellation Draco was the pole star for the Egyptians thousands of years ago. Vega will assume the title of pole star in about 12,000 years.[13] Polaris passes closest to the north celestial pole right around 2100 AD. This also means that the **equinoctial points** where the ecliptic crosses the celestial equator are also slowly moving as well. Live a typical 72-year life span and they'll have moved exactly one degree.

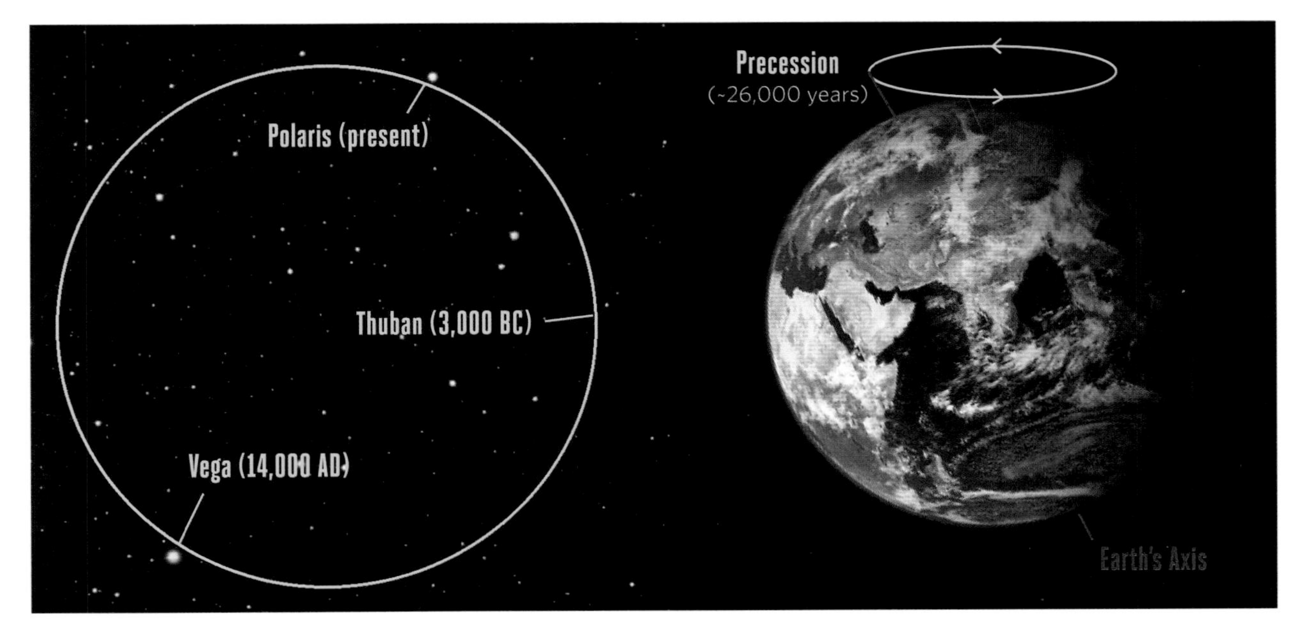

The 26,000-year Precession of the Equinoxes, as the Earth's axis wobbles like a top (right). The left diagram shows the position of the Earth's northern polar axis over the 26,000-year span.

This also means that the modern astronomical constellations are no longer aligned with the astrological houses assigned in Ptolemy's time thousands of years ago. The Sun actually passes through the thirteenth additional constellation of Ophiuchus the Serpent Bearer from November 29 to December 18, and the current offset means, for example, that Leos should be Cancers!

The stars themselves are also in motion. Our Sun, as mentioned, is 26,000 light-years from the core of our home Milky Way Galaxy, completing one revolution every quarter of a billion years. From the perspective of the Solar System, this motion (known as the **solar apex** or the **Apex of the Sun's Way**) is in the direction of the constellation Hercules[14] at 145 miles per second (233 km/s). All of the stars in the night sky belong to the Milky Way Galaxy, and though their motions may seem fast, space is a big place, and the patterns of the constellations will look pretty much the same on the day you're born as the day you die. They did, however, look slightly different to the ancient Greeks thousands of years ago, and they would have certainly looked very different to the dinosaurs millions of years ago, if they ever bothered to look up.

This apparent motion of stars over time is called **common proper motion**, often abbreviated as CPM. This is the sum of a star's motion toward or away from our Solar System, and across our line of sight. To give you some idea just how tiny CPM is, Barnard's Star—the star with the fastest common proper motion at a blazing 10.3 arcseconds per year[15]—crosses the apparent width of a Full Moon every 175 years.

Farther afield, we now know that many of those little fuzzy smudges in the sky are actually galaxies in their own right. If you're in the southern hemisphere, you have two fine examples of irregular galaxies that are nearby in cosmological terms but separate from our Milky Way: the Large and Small Magellanic Clouds. Up north, you may just be able to see the Andromeda Galaxy (Messier 31) on a clear, dark night.

Once astronomers were able to obtain spectra from individual galaxies, they realized something astounding: the absorption lines in their spectra were shifted to the red, meaning that on cosmological scales, every point in the Universe is rushing away from us, and from one another. When I first learned about this as a kid, I lay awake at night, in terror that the molecules in my body would suddenly fly apart. Don't worry too much about flying apart, however; on small scales, gravity and the strong nuclear force that binds the atoms of your body together are much more powerful than the expansion of the Universe. Though you won't see redshift at work during a casual glance at the night sky, we now know it's a reality in an expanding universe that's actually accelerating, meaning that billions of years from now, the last distant galaxy will recede beyond our cosmic horizon, and any other enlightened civilizations that arise after that will be certain that *their* home galaxy is alone and the sum total of the cosmos.

ORDERING MAGNITUDE

"Did you see Venus this morning?" a well-meaning coworker once asked me, knowing my interest in astronomy. "It was HUGE!" Invariably, people looking at the sky confuse apparent size with brightness. Astronomers use the term magnitude to describe the brightness of an object in the sky. The current scale[16] was first roughly formalized by Ptolemy in 140 AD, who ascribed a value of 1 to 6 from bright to faint for stars that he could see. The scale was calibrated by Norman Pogson in 1856 as the system we use today. The lower the number value in terms of magnitude, the brighter the object, and the modern-day magnitude scale runs into negative numbers. The magnitude scale is logarithmic, meaning that the increase or decrease in brightness is exponential. One full number change in brightness is equal to a 2.512 times change in brightness. A handy part of this seemingly arbitrary setup is that a change in five magnitudes is exactly a hundredfold change in brightness, a change in ten magnitudes is 100^{10} or 10,000, and a change of fifteen magnitudes (say, a first-quarter Moon versus the faintest naked eye stars you can see) is a millionfold change in brightness.

There is no theoretical limit to the magnitude scale at either end, but the stars you see in the night sky under the best conditions range from magnitude -1 to about +6. Binoculars will extend your light grasp down to about magnitude +10, and a good telescope can extend this light grasp further still, down to about +15. In the opposite direction, Sirius is the brightest star in the sky at -1.46, and the planet Venus can shine at magnitude -4 at greatest brilliancy, bright enough to cast shadows. The Full Moon is brighter still at -13, and the Sun shines at a dazzling magnitude -26.[17]

When we're describing how bright objects appear in the night sky, we're talking about **apparent magnitude**. Light falls off inversely with the square of the distance from the observer; move a candle twice as far away, and it'll appear a fourth as bright. The same is true for stars: some are relatively nearby, some are far away. To calibrate the situation, astronomers use a system of **absolute magnitude**, which imagines how bright a star would be if it were placed 10 parsecs (32.6 light-years) distant. Put our Sun (a garden-variety yellow dwarf star) 10 parsecs away, and it would shine at a paltry magnitude +4.7, barely visible to the naked eye.[18] Move a distant quasar like 3C 273 to the same distance—an ill-advised exercise, as you'd fry the Earth to a crisp—and it would outshine the Sun at a brilliant magnitude -27.

Finally, the nature of the light source can make a huge difference as to what you see with the naked eye and the eyepiece. Stars, though they're massive close up, are actually distant and visually tiny, so small that they appear as dimensionless points of light. Planets are very nearly so, as they have a diameter measured in scant arcseconds. This means that the quoted brightness in magnitude is radiating from one point. In the case of extended objects such as comets or diffuse nebulae, however, all that brightness is smeared out over an extended area, making the apparent visual brightness of the object fainter than the stated magnitude. This is known as **surface brightness**, and is usually measured per square arcsecond versus luminosity.

Looking through the thick layer of the atmosphere close to the horizon can also dim an object down a magnitude or two, in a phenomenon known as **atmospheric extinction**.

"DID YOU SEE VENUS THIS MORNING?"

a well-meaning coworker once asked me, knowing my interest in astronomy. "It was HUGE!"

SELECTING A PRIME OBSERVING SITE

Where you select to observe the sky is crucial to a successful deep sky observing experience. Unfortunately, few of us (excluding my adult self) enjoy the truly dark sky speckled with stars that was visible from my doorstep in northern Maine as a kid. Many observers now drive hours from home to find the same sort of sky. But don't despair; we've carried out successful impromptu sidewalk star parties from the brightest urban streets of downtown Tampa or the strip in Las Vegas (arguably the most light-polluted observing site in the world), so the odds are you can see something in the sky from your own rooftop or backyard.

First, you want to select a site that is blocked in from street lights and light pollution. Hills or unlit buildings make great little shadow islands in a sea of urban brightness. Be careful, however, that your site isn't hemmed in from the horizon, or your view will be restricted to objects directly overhead. Beaches and large farm fields make the very best viewing sites, providing long, expansive views of the sky.

Again, light pollution—or the absence of it—will make all the difference, especially when it comes to hunting for extremely faint deep sky objects. Our eyes need time to dark adapt to faint light, and often, a faint, fuzzy nebula will pop in and out of the scene just out of the center of the eye's view, using a trick known as **averted vision**. This works because the cells around the fovea of the central eye are slightly more sensitive to dimmer light.

People lacking power in South Florida during Hurricane Andrew in 1992 marveled at the view of the nighttime Milky Way spanning the sky over usually overlit suburbs. Friends who deployed to Afghanistan often also remark how truly dark and pristine their skies are over there—military bases in theater are often unlit, so as to not draw sniper and artillery fire—producing a pristine, dark nighttime sky unknown to the troops back home.

USING DESKTOP PLANETARIUM SOFTWARE TO UNDERSTAND THE NIGHT SKY

It has become a familiar sight at modern star parties. Glowing smartphone screens waving at the night like lighters at a concert, searching for planets and deep sky objects.

There are plenty of planetarium apps out there, which use your phone's GPS to display a representation of the sky overhead. Google Sky is a great free utility, and we routinely use Heavens-Above's free lite version to track satellites overhead in the field.

A good desktop planetarium program is also an indispensable tool for planning a night's observing session. You can also clear away the two biggest obstacles we have for Earthbound observing—the ground and the atmosphere—to get an idealized view of the sky.

Powerful programs such as Starry Night Pro and Redshift can put the sky at your fingertips. One great free resource is Stellarium. Using Stellarium, you can explore the sky from your location, use overlays of equatorial and alt-azimuth grids, and see what the sky looks like from your locale. Amateurs even use ASCOM drivers[19] to control telescopes using planetarium programs.

Planetarium programs and apps are a great way to familiarize yourself with the night sky, and give you a mental map of our vantage point in space and time. They also provide the guilty pleasure of giving us views of what we *can't* see: Who wouldn't love to gaze at Saturn as seen from its moon Iapetus (with the proper space-suited protection, of course) or witness the Moon occulting Saturn during a total lunar eclipse on July 26, 2344?

ACTIVITY: FINDING DARK SKIES

How dark is your sky? In Chapter 11, page 208, we'll talk about how to gauge the darkness (or lack thereof) of the sky from your backyard.

David Lorenz also maintains a regularly updated *World Atlas of Artificial Sky Brightness* with zoomable maps created using the very latest in Earth-observing satellite data. The results are sobering, to say the least: http://www.inquinamentoluminoso.it/worldatlas/pages/fig1.htm

I like to access the map as an overlay for my current location using Clear Sky Chart: http://www.cleardarksky.com/

The International Dark Sky Association (IDA) maintains an extensive survey of dark sky sites worldwide: http://www.darksky.org/idsp/

Fourteen Selected International Dark Sky Parks Worldwide

- Big Cypress National Preserve (U.S.)
- Bodmin Moor Dark Sky Landscape (England)
- Craters of the Moon National Monument (U.S.)
- Death Valley National Park (U.S.)
- Elan Valley Estate (Wales)
- Flagstaff Area National Monuments (U.S.)
- Galloway Forest Park (Scotland)
- Joshua Tree National Park (U.S.)
- Lauwersmeer National Park (The Netherlands)
- Ramon Crater (Israel)
- Warrumbungle National Park (Australia)
- Waterton-Glacier International Peace Park (U.S./Canada)
- Yeongyang Firefly Eco Park (South Korea)
- Zselic National Landscape Protection Area (Hungary)

Finally, don't forget to search out advice from local astronomy clubs: dark sky observing sites are to amateur astronomers what secret fishing holes are to fishermen.

Don't forget to dress for the dark and a session of observing as well. When you work out, it's easy to dress light even in cold weather, as your body heats up like a furnace. Astronomy, however, often involves long stretches of standing relatively still, and you can radiate out that precious heat in a hurry. But don't forego cold weather observing, as it's often when the skies are settled down and the views and seeing are tack sharp and at their very best. We've observed down to -40°C outside of Fairbanks, Alaska, where focusers freeze and even a brief icy breath can coat an eyepiece.

Dress in layers and wear two-layered gloves in extreme cold, so you can remove one and keep another on for fine motor manipulation of cameras or equipment. A warm thermos of tea or soup also helps. It's also handy to keep a small propane space heater nearby to warm batteries or hands over, though keep it far enough away from the telescope to avoid creating local turbulence. Finally, wear insulated boots, as hard-soled shoes will conduct body heat out to cold concrete in a hurry.

A frozen DSLR camera with frost build up. . . don't let this happen to you or your gear!

ACTIVITY: MEASURING THE SKY

My challenge to you is to try to forego the apps and tech while out actually observing, at least early on. Instead, use them during the daytime to learn what will be visible tonight, then try and make sense out of the sky overhead after dusk. Watch a telescope operator at a star party with years of experience and how they simply just *know* how to find the M13 globular in the Keystone of Hercules as easily as you know how to find your way back home every night. This is an exercise in muscle memory, as you come to know the sky like an old friend after decades of experience. I promise you will never forget how to find your way in the night sky.

Most planetarium programs have a measuring tool, handy for making sense of the scale of a view in degrees of arc. Below is a handy way to simulate this very application tonight, using your own body.

- Standing upright, extend your open palm out to about an arm's length.
- Look at your outstretched hand against the sky. The span of your palm from the bottom of your hand to the top of your thumb is roughly about 10 degrees of arc, with each finger's width representing about 2 degrees.
- Remember the scene in *Apollo 13* where Tom Hanks covers up the Moon with his outstretched thumb? You too can smudge Earth's natural satellite out of the sky, proving its apparent size is much smaller than most folks think.

I've used this method of hand measurement to locate planets against the dusk or dawn sky, or point my binoculars to a patch of sky where a newly discovered comet ought to be. A bright twilight or light-polluted sky might provide few landmarks, and you can use your very own hand measurement tool to springboard from the local horizon, the Moon, or a bright star to find your target. And the neat thing about this method is that the human body is roughly proportionate, meaning that the 10 degrees for an outstretched hand rule is true for folks young and old, large and small.

It's almost as if we are built to measure the cosmos.

Measuring angles in the sky with an outstretched hand.

ASTRONOMY GEAR AND TECH TALK

The tools of astronomers and how to use a telescope.

"Each star had cost an effort. For each there had been planning, watching, and anticipation. Each one recalled to me a place, a time, a season. Each one now has a personality. The stars, in short, had become my stars."

— *Leslie Peltier*, Starlight Nights: The Adventures of a Star Gazer

PICKING THE PERFECT TELESCOPE

ACTIVITIES

- Building a simple Newtonian refractor for under $50 (page 36)
- Building an observatory for under $500 (page 44)

USING A TELESCOPE IS LIKE LEARNING TO PLAY A MUSICAL INSTRUMENT.

Sooner or later, we all want one.

Maybe you've just received a telescope as a gift, and have no idea how to use it. It's a common sight at public star parties: Someone approaches me with a brand-new telescope, asking me to show them how to use it as dusk falls. I usually oblige if they agree to set up next to me, hoping all of the parts are available and that I can run through the quirks of this particular telescope before it gets dark.

First, I'll let you in on a few secrets: Using a telescope is like learning to play a musical instrument. Like learning to read sheet music and play chords, there's a language to learn, and it takes practice and patience. I'm often asked how I know where the Ring Nebula or the Andromeda Galaxy is, but after showing them off at public star parties for years, slewing the telescope to them is part of automatic memory, like playing the guitar or knowing how to drive home.

It's for this reason I like to steer prospective telescope users away from sophisticated GoTo instruments. These have a place, and it's handy to have a computer guide to help find and track a faint obscure object. But using a simple manual mount like a Dobsonian gives you a good, intuitive feel for learning the night sky.

MAGNIFICATION VERSUS APERTURE

"Delivers x1,000 magnification!" the breathless announcement says on the outside of the box, adorned with glorious full-color images of Saturn and the Orion Nebula. This is the typical pitch for a department store refractor, usually a small 60 mm instrument. Here's another secret: Magnification isn't as important as you might think. Such an instrument achieves such a high magnification by use of a narrow field eyepiece and a long tube Barlow lens. However, such an instrument also suffers from a narrow field of view. Hold the empty hollow tube from a pen up to the sky, look through it with one eye, and try to find a selected star. That's the situation you face with one of those cheap department-store telescopes. Often, such instruments are little more than toys, with plastic lenses and flimsy mounts. Plus, a high-magnification view also magnifies turbulence in the atmosphere, giving you a murky scene that refuses to snap into focus.

Left corner: Startrails in a time exposure over Glendoll, Scotland, documenting the nightly apparent path of the sky as the Earth rotates.

Aperture, or the size of the mirror or lens capturing the incoming light, is much more important. The aperture of your pupil when it's dark adapted, for example, is about 8 mm. The eye isn't a splendid optical device, but considering that nature and evolution had only water and jelly to work with, it's amazing that it works at all. Think of the aperture of a telescope as extending and widening that eye opening to a pupil several inches wide or more. This, in turn, extends your light grasp, allowing you to see fainter objects at a higher resolution.

Magnification is also a function of **focal length**, which is the distance light travels from the primary collector to the eyepiece. For example, an 8-inch (20-cm) telescope might have a 1,000 mm focal length, so using a standard 25 mm eyepiece yields 1,000/25 = 40 times magnification. I actually use a 42 mm eyepiece with a similar setup about 90 percent of the time . . . such a low-power 24x magnification setup provides a generous low-power field of view, making sweeping up faint comets or glorious globulars a snap.

For example: A wide field of view might allow you to take in the entire disk of the Moon or a wide extended object such as the Pleiades star cluster . . . but high magnification is handy for small planetary nebulae, or splitting close double stars.

A large aperture-reflecting telescope in action.

APERTURE FEVER

The primary danger at any star party isn't tumbling off a ladder in the dark. It's actually the threat of looking through someone's ginormous 18-inch (46-cm) Dobsonian and coveting that very view for yourself.

Welcome to aperture fever, a very real occupational hazard for anyone starting out in astronomy. No matter how large a light bucket you have, there's always one out there that's bigger. We've seen up to 24-inch (61-cm) reflectors at star parties. Probably the biggest amateur telescope ever conceived is the 70-inch (178-cm) monster built by Mike Clements out in Utah.[1] True story: He actually built this telescope around a surplus spy satellite mirror he picked up at a government auction in 2005.

Don't forget, though, unless you construct an observatory to permanently house that light bucket, you'll have to transport all of that weight. An amateur astronomer has a slightly different world perspective when it comes to the usual familial obligations of wants and needs. I've known amateur astronomers who actually select their family vehicle on the criterion of whether it will carry their telescope!

For years, my star party go-to telescope has been a fork-mounted Schmidt-Cassegrain 8-inch (20-cm) telescope. It's an easy setup, and it has taken enough abuse to be well broken in. The clock drive isn't quite accurate enough for photography, but an 8-inch (20-cm) mirror still delivers generous views of the sky.

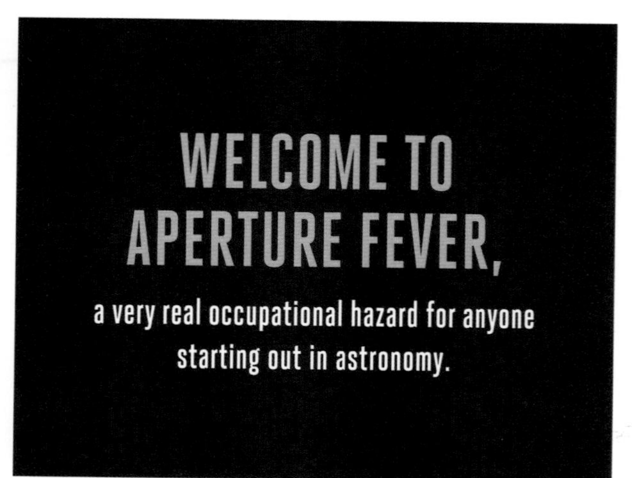

Three types of telescope mounts: a Dobsonian on a "lazy Susan"–type mount (left), a large reflector on a fork mount (the 1-metre reflector at Embry-Riddle University in Daytona Beach, Florida) (right), and a typical German equatorial mount (center).

A MOUNTAIN OF MOUNTS

Ninety percent of your happiness (and frustration) will come from the stability of the telescope's mount . . . or lack thereof. Here's a good, quick test: give the tube of your telescope a good whack. Not a knock-you-down, smack-out-of-left-field hit, mind you, but the kind of slap you'd use to swat a mosquito with. Does the view settle down in about three seconds or so? If not, you're saddled with a wobbly, trembling mount that will flex and bend with the whims of the wind, the touch of a hand, or people just walking by.

Telescope mounts come in two different types, reflecting two different coordinate systems in the sky: **alt-azimuth** and **equatorial**.

Alt-azimuth mounts are the simplest, with one axis pointing straight up at the zenith for left-right rotation in azimuth, and another horizontal axis to move the telescope in altitude or elevation. **Dobsonians** are alt-azimuth mounted, and provide great stability by placing the telescope's center of gravity down low to the ground. This type of mount is also very intuitive, and provides a user with an easy way to simply push and pull the telescope toward the target. The main disadvantage of an alt-azimuth mount is that you have to keep nudging the object back into the field of view as the Earth rotates, which is also a hindrance for long-exposure astrophotography.

Enter the **equatorial mount**. This type of telescope mount tracks objects in right ascension, with one axis aimed at the celestial pole. Working with a clock drive, this means that you can aim a polar-aligned equatorial mount at an object, and it will track an object as the Earth rotates, great for showing off the view to a long line of curious viewers.

Equatorial telescopes come in two varieties: fork-mounted and German equatorial. **Fork** or **horseshoe mounts** are simple and intuitive to use, and can be thought of simply as an alt-azimuth mount tipped back so that one axis is pointed at the north or south rotational pole, depending on your hemisphere.

German equatorial mounts are a little trickier to use. Sometimes called a T-mount, a German equatorial mount (GEM) has one axis stacked on the other. The chief benefit to this configuration is its compact design, ideal for small telescopes; the design also has no blind spots, and can reach targets at high declination or near the zenith. Typically, though, this type of mount needs to be balanced before use with large counterweights offsetting the tube of the telescope . . . nearly every amateur astronomer has a story to tell about releasing the axis clamps on their German equatorial mount in the dark, only to have the telescope come crashing down because they failed to install the counterweights! This can be an expensive physics experiment to conduct in the dark, proving gravity does indeed work. German equatorial mounts also require a meridian flip maneuver as they track objects across the north-to-south line, plus pointing and positioning them can be counterintuitive and takes some practice.

Don't overlook the sturdiness of the mount when purchasing a telescope. Be prepared to spend as much on your mount as you will on the telescope.

EYEING EYEPIECES

All telescopes use eyepieces for the final collection point to get light into the eye. These days, most standard eyepiece barrels are 1¼ inches (3 cm). This standard actually came about because the telescope maker John A. Brashear started using plumbing pipe fittings for eyepieces back in 1890![2] Sometimes, larger, more sophisticated eyepieces might use 2-inch (5-cm) barrels. There are high-quality, wide-field eyepieces offering spacewalk-style views out there that cost more than many telescopes. Very occasionally, you might come across older 0.965-inch (2.5-cm) barreled eyepieces on Japanese-made Tasco, Jason, or Unitron refractors from the 1960s and 1970s.

There is a bewildering array of telescope eyepieces out there today. Early two-element design eyepieces gave way to complex multiple-element designs such as the Nagler or Plössl eyepieces, which are popular today. These come listed with their own separate focal length in millimeters. A good wide-angle eyepiece offers a generous field of view for a floating-out-in-space sort of visual experience. These are a great upgrade but again, be warned: You can spend more on a high-end eyepiece than a telescope! That being said, it's worth having at least one low-, medium-, and high-powered eyepiece in your kit, as well as a 2x Barlow lens to really crank up the magnification. For star parties, consider investing in a single zoom eyepiece, so you can skip continually changing out eyepieces for long lines of viewers.[3]

RESOLVING RESOLUTION

To most people, a resolution is something they make but never keep around New Year's each year. In astronomy and photography, **resolution** refers to the tiniest details you can see, a term photographers often call the graininess of an image. The theoretical resolution of a telescope is described by the **Dawes Limit** and the following equation: $r = 11.6/d$, where d is the telescope aperture in centimeters, and r is the best possible resolution in arcseconds. Thus, a 20-cm reflector with a 20-cm aperture breaks down as 11.6/20 = 0.58 arcseconds. Remember, though, this is a theoretical limit, and bad seeing, turbulence, and atmospheric transparency all conspire to assure you never fully reach this ideal limit.[4]

Fun with the Dawes Limit: A typical dark-adapted eye might have a maximum aperture of 5 mm or 0.2 arcseconds. The expected resolution of your "Mk-1 eyeballs" is therefore, 11.6/0.5 = 30 arcseconds, or just under the aforementioned 1 minute of arc. This is an ideal theoretical limit, under perfect conditions.

To most people, a resolution is something they make but never keep around New Year's each year.

CONSIDERING BINOCULARS

Chances are you might have something lying around the house that you can use to examine the sky with tonight: binoculars. A good set of binoculars is indispensable to backyard astronomy, and offer up a generous true field of view. And while you can spend the cost of a good used car on a set of high-end binoculars, a set of decent 7x50 binoculars typically used for hunting or birding need not cost more than $100, and are also suitable for astronomy. Binoculars offer a quick magnified view of the sky, with no setup or alignment necessary. I started into astronomy using my dad's 7x50 binoculars to see craters on the Moon, nebulae, and the moons of Jupiter. Plus, most binoculars are nearly indestructible, and they easily fit in an airline carry-on and can come along on any backpacking expedition. I've owned telescopes big and small, but I'll admit that we're still using binoculars 90 percent of the time . . . though I'll also make a small concession that ours are Canon 15x45 image-stabilized binoculars, and they were expensive. Image-stabilized binoculars are dangerously addicting, in that once you're used to using them, you don't want to go back to plain old ordinary non-stabilized binoculars. I now find myself reaching for the nonexistent stabilizing button when handed a pair of ordinary binoculars.

Binocular specifications are described using two numbers: "7x50" means 7x magnification, with a 50 mm aperture. **Diopter spacing** states the distance between the two eyepieces, usually adjustable by hinging the two halves of the binoculars apart or closer together. One eyepiece on most binoculars is adjustable, so you just need to bring the fixed one into focus, then adjust the other from that point. This also means, however, that it's often difficult to share the view with binoculars, as everyone has a slightly different pupil spacing and acuity difference between their left and right eye, meaning they have to readjust the binoculars each time they're passed back and forth.

Looking through binoculars is very different than looking through a telescope. Our brain stacks the images from each eye using a technique known as **binocular summation**, so the result is better than the addition of each individual eye.

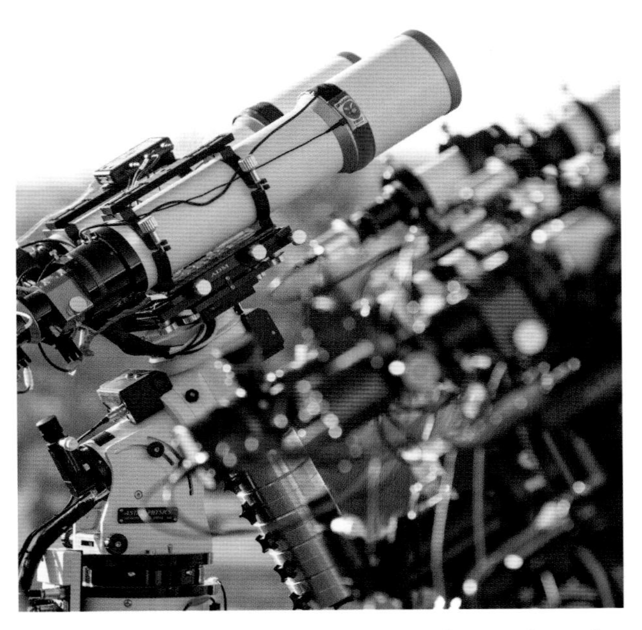

A "flock of refractors" is known as a (?) A typical refracting telescope in action.

TYPES OF TELESCOPES

The telescope is an icon of astronomy. From Galileo's first crude instrument to the Hubble and James Webb Space Telescopes, telescopes have become an instrument designed to collect and focus light for either a human eyeball, camera, or imaging package at the opposite end.

There are two basic types of telescopes: refractors and reflectors.

Refractors use lenses exclusively. This is also the type of telescope the public is familiar with: the simple sort of collapsible seaman's draw tube, with one large lens in front of the other. Their simple construction also means that refractors were also the first type of telescope that sixteenth-century Dutch opticians hit upon, the same sort of

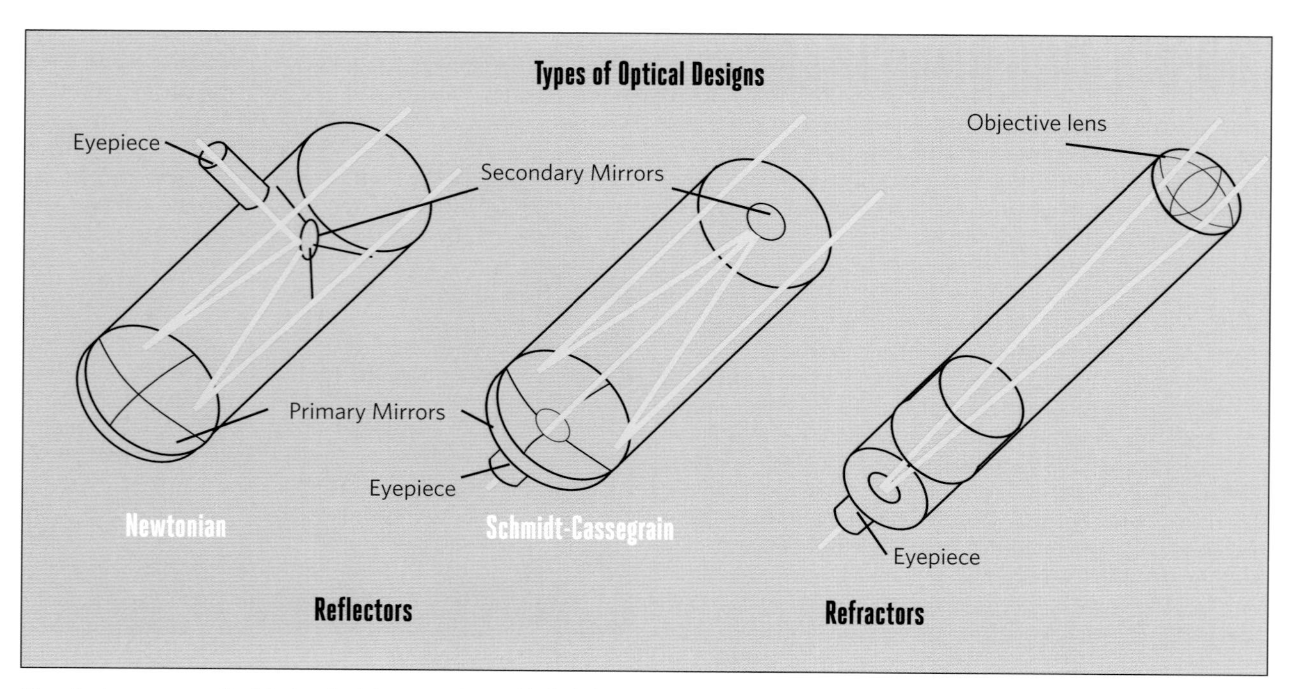

Three basic, common optical designs for telescopes, along with the light path (in yellow) from the telescope aperture to the eyepiece. Nearly every telescope design is a variant of these three basic configurations.

crude design that Galileo aimed at the night sky and steadily improved upon. Galileo also came across the bane of refractors known as **chromatic aberration**. Often referred to as color fringing, this is caused by the curvature of light reaching focus at different points,[5] giving objects an unwanted rainbow hue. One easy way to overcome this was to use thin lenses, which necessitated absurdly long focal lengths seen in the telescopes of the seventeenth and eighteenth centuries. It wasn't until 1733 that Chester Moore Hall hit upon the idea of compound crown and flint lenses to defeat this problem, a system still used in telescopes today.[6]

A refractor offers a compact, portable design, relatively free of maintenance. The chief drawback, however, is the size: refractors scale up pretty quickly in terms of size and weight, making the largest amateur refractors just 8 inches (20 cm) in aperture. The largest functional refractor is the Yerkes 40-inch (102-cm) refractor in Wisconsin built in 1897.[7] This grand old telescope will probably forever hold the title of the world's largest.

Large, modern instruments tend to be reflectors, the second type of telescope that makes use of a large curved primary mirror to collect light. Isaac Newton constructed the first practical **reflecting telescope**, a small 2-inch (50-mm) aperture instrument[8] that was focused by adjusting a screw on the tube's end. Today, Newtonian reflectors are a common design seen at many star parties, consisting of a concave primary,[9] a tilted secondary mirror, and an eyepiece mounted on the side of the tube.

Early mirrors were made of a polished metal known as **speculum**, an amalgam of copper, nickel, and arsenic.[10] These early mirrors tarnished quickly and required constant attention. Advances in mirror technology had to wait until the early nineteenth century,[11] when the process of silvering mirrors was perfected. Early silvering techniques used a mixture of tin and mercury, then later, silver. Today, mirrors are generally coated with aluminum (a process known as sputtering) to achieve better than 95-percent reflectivity.

A **Newtonian reflector** is the easiest telescope for an amateur astronomer to build, as it consists simply of a parabolic primary mirror and a secondary tilted mirror that sends the light to an eyepiece mounted on the side of the telescope's tube. Prior to the advent of mass-market telescopes in the 1960s and after, homemade Newtonians were the *only* way amateurs could gain access to a large-aperture telescope.

Another design, known as the **Schmidt-Cassegrain reflector**, has also gained popularity in the past few decades. This is a hybrid design combining mirrors and lenses in what's known as a **catadioptric** configuration. The primary mirror has a concave, spherical curve for the primary and secondary used in a Schmidt camera (as opposed to a classical Cassegrain telescope, which has a concave, parabolic primary mirror and a convex, hyperbolic secondary). A Schmidt-Cassegrain telescope uses a large glass corrector plate to overcome spherical aberration. The advantage for this design is a large aperture that's relatively portable: I can easily sling my 8-inch (20-cm) Schmidt-Cassegrain telescope (SCT) and tripod mount in the back of my Fiat 500 (a tiny Euro car) for a night of observing in the field.

An SCT also requires very little mirror adjustment (known as **collimation**). Some Newtonian telescopes require recollimation *every* time they're assembled in the field for a night's observing. I've pulled our trusty SCT telescope out of its case after shipping it around the world to find the mirrors in near-perfect alignment. You can tell if your mirrors are out of alignment with a simple test: Find a bright star, center it in the eyepiece field of view, then deliberately throw it out of focus. It should look like a silvery doughnut, with a doughnut hole dead center. That's the shadow of the secondary, and if it's off to one side or the doughnut looks squished, the mirrors are out of alignment. With a Newtonian reflector, the primary and secondary mirrors have three adjustment screws each—an SCT only requires adjustment to the secondary, as the eyepiece tube goes through the hole in the center of the mirror and is fixed perpendicular to the primary mirror.

Of course, refractor owners never have to adjust their telescope alignment at all, as the lenses simply stay fixed in their cell mounts.

Stranger designs for telescopes exist, mostly variants of reflector telescopes. Some amateur designs incorporate tertiary mirrors and folded optical paths in an effort to overcome the light lost by placing the secondary mirror in front of the primary. Two variants of the SCT design have made their way onto the commercial market in recent years: the **Ritchey-Chrétien** reflector, a high-end optical design, and the **Maksutov-Cassegrain** ("mak") design similar to an SCT, which uses a meniscus corrector lens on the curved corrector plate for a very compact design.

Which sort of telescope is right for you? I'd say that the right telescope is the one you use most. If you travel a lot (like me), you want a short tube refractor or small Maksutov-Cassegrain with a stable but portable mount. Do you want a scope simply for casual observing? Then you can't go wrong with the simplicity of a Dobsonian light bucket reflector, though if you think you're going to get seriously into astrophotography, you want a sturdy German equatorial or fork mount capable of tracking the sky during long exposures. Likewise, doing sidewalk astronomy with an alt-azimuth mounted Dobsonian means that every 5 minutes or so, you'll have to nudge the target object back into view, a somewhat tedious prospect for long lines of observers. A roughly polar-aligned scope can still track an object for about an hour, fine for casual public observing.

So as you can see, telescopes are like automobiles, in that there are trade-offs between designs, and the perfect auto is often based on the application it's used for. A sports car might be great for going from 0–80 mph (0–130 kph) in seconds, but a terrible choice for picking up furniture from IKEA or the soccer team from school.

We've built telescopes and owned reflectors and refractors of various makes and models, but our trusty 8-inch (20-cm) SCT seems to be our mainstay, a good combination of portability, a generous aperture, and a sturdy equatorial mount.

WHICH SORT OF TELESCOPE IS RIGHT FOR YOU?

I'd say that the right telescope is the one you use most.

ACTIVITY: BUILDING A SIMPLE NEWTONIAN REFRACTOR FOR UNDER $50

Why build a telescope? It's a good question. After all, in this age of mass-market telescopes, good high-quality, large-aperture telescopes have never been cheaper. Until about the 1970s, large, serious telescopes for astronomy were unavailable for purchase, and a 6-inch (15-cm) reflector was considered a large telescope. If you wanted a larger instrument, you built it, mirror and all.

Still, there's an argument to be made for building your own telescope, even today. Nothing gives you an intimate appreciation for just how something works like building it yourself. Many master mirror makers can still reach a quality nearing military specifications, better than most factory-made telescopes. And if you want to incorporate something exotic, such as a folded optical design or a tertiary mirror with a Nasmyth focus, then your option is still pretty much to build it yourself.

I'd like now to present a third hybrid option available to amateur telescope makers: simply assembling a quality telescope from parts.

I undertook just such a challenge years back. I had the idea of using a 6-inch (15-cm) stovepipe (a common accessory in chilly northern Maine) to make a simple Newtonian reflector. The challenge was to build this old stovepipe tube into a functioning astronomical telescope . . . for $50.

Step 1: Gathering the Parts

The primary mirror and the secondary mirror are the most complex parts of any reflecting telescope. These days, complete factory mirror kits can be found on eBay, Craigslist, and Amazon. Often, sellers acquire these kits in bulk from wholesalers. For example, we acquired a 5.5-inch (14-cm) mirror kit complete with a secondary and mounting brackets from an eBay seller for the princely sum of $19.99. It was listed as a "factory reject" due to slight blemishing, but still works fine for casual observing.

Supplies and Tools Needed

- One 6-inch (15-cm) diameter, 48-inch (14-cm)-long section of stovepipe
- Electric drill
- One 5.5-inch (14-cm) mirror kit, complete with the secondary mirror and attaching hardware (a more ambitious project would be to grind your own mirror)
- Two binder clips and attaching hardware for the laser-pointer mount (optional, if you're mounting a Telrad finder on the tube as well)
- Short 2-inch (5-cm) spacer-length section of 3-cm PVC pipe and/or copper piping (I ended up using both to reach focus) and four small L-brackets plus four screws and nuts for the eyepiece holder
- Phillips screwdriver
- Crescent wrench
- Small $5 handle with attaching hardware
- Grease pencil
- Super Glue
- Mirror collimation tool (optional), such as a laser eyepiece collimator (costs about $25)
- Set of jeweler's screwdrivers

Step 2: Assembly of the Telescope

Assembly is as simple as drilling screw holes and mounting the primary and secondary in the tube. The mirror cell will have alignment holes to guide you for drilling the three ¼-inch (6-mm) mounting holes on the stovepipe, while the secondary vanes will need three ¼-inch (6-mm) holes spaced 120 degrees apart. A crucial step is making sure the secondary is in the right place to reach focus; with an f/8 mirror, the prime focus point is 5½ inches (14 cm) x 8 = 44 inches (112 cm) from the mirror's center. See page 34, Types of Optical Designs, for the light path of a typical Newtonian reflector.

Step 3: Assembly of Extra Hardware

Assemble the focuser from parts, using PVC pipe, screws, and right-angle brackets (listed above). It is a simple helical focuser, meaning you slide the eyepiece to focus, then turn the set screw tight. The focusing set screw is nothing more than a ¼-inch (6-mm) screw with a washer superglued in the screw slot, inserted into a ¼-inch (6-mm) hole drilled into the PVC or copper pipe. The plumbing section of the hardware store is a great place for telescope makers, with tube fixtures compatible with the 1 ¼-inch (3-cm) standard telescope eyepiece barrels of today.

Step 4: Optical Alignment

The final step is collimation. A laser collimator is a quick and handy tool to have, though we used homemade crosshairs made of wires over the eyepiece tube for this one. Many primary mirrors have a small black dot in their center to aid with this adjustment, or you can use a temporary doughnut-shaped reinforcement label sticker for this step. Both the secondary and primary mirrors have three set screws to adjust, which will push and pull the mirror in and out of alignment. With the eyepiece out of the eyepiece holder, you will slowly tweak the set screw adjustment on each mirror until the secondary and the black dot on the center of the primary mirror are aligned like a bulls-eye. (See the image on page 41 for doing this with a laser collimator.) You may need a friend to do the primary mirror adjustments while you look at the lineup through the eyepiece holder.

Use a small jeweler's Phillips-head screwdriver to adjust the secondary mirror set screws, but make sure you do this with the telescope tube horizontal and not in the vertical position—you don't want to drop the screwdriver down the tube onto the primary mirror!

Step 5: Adding a Telrad and a Sturdy Mount

Finally, add a set of metal binder clips for a laser pointer, a handle, and a Telrad mount for the finder. We initially used clamps with a ¼-inch (6-mm) universal mount to attach the scope to a camera tripod, though this turned out to be a bit wobbly. Years later, we built a simple but sturdy Dobsonian mount out of plywood, two tea tins for bearings, and bungee cords. This final configuration with the stovepipe scope on a simple Dobsonian mount is pictured and labeled below.

Step 6: First Night Out

This sort of found telescope built with surplus parts is a fun weekend project. With a medium-power eyepiece installed, the First Quarter Moon filled the view, but it was wide enough to easily sweep up deep sky objects. It also provided some detail at low power on planetary surfaces, such as Jupiter and Saturn. And, it was ready just in time, as Stovepipe Scope saw first light during the historic outburst of Comet 17/P Holmes in late October 2007.[12]

A simple 5½-inch (14-cm) Newtonian reflecting telescope on a Dobsonian mount, built around a 6-inch (15-cm) stovepipe.

NIGHT ZERO

Learning to use a telescope, like learning to play a musical instrument, means that competence comes with practice. Likewise, simplicity is always best in the beginning. For your first telescope, bypassing GoTo driven mounts promising a "database of 10,000 objects!" is a good idea. Better to learn your way around the sky first, as an intuitive way to intimately know the sky. Also, going for a simple alt-azimuth mounted refractor or a Dobsonian reflector that can simply be manually pushed and pulled toward a target means no troubleshooting problem mounts and complex drive systems in the dark. GoTos have their place—I've used them to track down faint, unfamiliar fuzzies and new comets with great success. It's just that they often have their own bugs to work out (like when your telescope barrel aims stubbornly at the ground) and they can add to the frustration for a first-time user when they don't properly put a target in the field of view on the first (or maybe seventeenth) try. Newer GPS-enabled telescopes fare better in this regard, but we've yet to see a GoTo telescope that is as simple to use as a point-and-shoot camera.

Likewise, while electric-drive motors are handy, we prefer to have the option of manual control, as we can simply keep observing when the batteries die. (This is the other reason we've kept our 1990s-era Celestron SCT telescope around.)

FINDING FINDERSCOPES

This brings us to our next often-overlooked feature that will either be the source of joy or consternation: the finderscope. A finderscope gives you an initial wide-angle view to (hopefully) center on a target, which will then allow you to sweep it up in the telescope's eyepiece field of view. Some stock finderscopes magnify the view slightly, while others offer a simple 1x true view of the sky, unmagnified. An illuminated Telrad or red dot 1x finder is a great addition to a telescope. These work much like the heads-up display on an aircraft, and simply paint the red-glowing reticle on a glass finder over the sky. These are simple, quick, and easy to use. Once I discovered the Telrad, I installed a base mount for mine on every telescope tube I own.

Finders always have a three-screw adjustment. The goal is to set it up to aim (or boresight) precisely where the telescope aims. It's best to do this initial adjustment in the daytime, while familiarizing yourself with your new telescope at the same time. The dark of night is a terrible time to try to figure out a strange telescope. Plus, you're wasting valuable observing time. Simply aim the telescope at a distant object, such as the top of a tree or a light post. Center it in the main eyepiece, then tweak the finderscope set screws until the same object is centered as well. Your finderscope is now aligned and ready for a night's worth of observing.

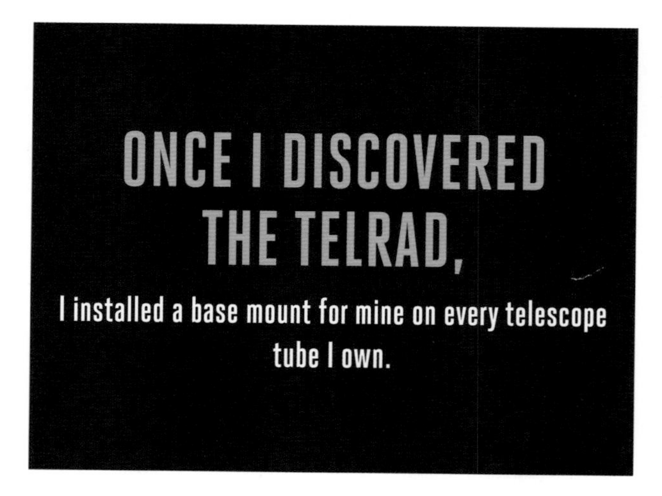

A green laser pointer pierces the night sky over a star party at Ash Meadows National Wildlife Refuge, Nevada.

OTHER USEFUL GEAR

But before you take that telescope out for its first night under the stars, there are two other pieces of gear that you might want to consider bringing along.

The first is a red light. A red flashlight will help preserve your night vision. It takes time for your eyes to become dark adapted, and once you've been blinded by headlights or a white flashlight, it can take up to an hour to really get that adaptation back. Plus, if you're out with a group for a serious night of astronomical deep sky observing, you won't win any friends by shining a bright white light around. You need to see what you're doing and be able to read charts and so on, and a red light will aid in this. Likewise, most astronomy planetarium apps have red-filter settings for use in the field. We've seen red plastic (think car taillight covers), red cellophane, and red nail polish used to turn ordinary flashlights into red lights for astronomy. And speaking of cars, be sure to locate how to kill your car's dome light (or physically cover it) *before* dark; otherwise, you'll be blinding everybody every time you duck into your car to retrieve something.

Red headlamps are our favorite, as you can use them hands-free. These are common with military and special operations types these days, and are often available at local hardware stores. Make sure the red aspect is truly red; some advertised red lights actually look more orange to our eye. Also, be sure you can switch the light from off to red to off again without having to dial through the blinding white light settings (most military-grade lights incorporate this feature).

The next piece of gear you should consider for your astronomical inventory is a green laser. These are great, especially for astronomy outreach. True story: My wife bought me a green laser for Christmas after she got tired of me saying "look over there" and gesturing in the dark. A 5-milliwatt (mW) handheld laser is about the size of a ballpoint pen and shines via collimated light reflected off of dust and debris in the air, making it a great pointer to aim about the sky and handy at public star parties. Green lights work opposite to red lights in terms of visibility versus preserving night vision, but this isn't really an issue at public star parties, as the "no blinding white light" rule is usually out the window when dealing with the general public. I've also seen green lasers used as finders, with one user aiming the laser at the target in the sky, guiding the person pointing the telescope.

That being said, green lasers aren't toys. Never aim them at a person, an aircraft, or an automobile, as they can blind someone at close range. In fact, I would go so far as to say you should *never* aim them anywhere but toward the sky. You never know if someone might be standing off a ways in the dark. Not only are some lasers more powerful than the stated 5-mW average, but military-grade lasers are also available on the market today, rated toward a powerful 1 watt.

Plus, kids (and many adults) are fascinated with handheld lasers, often more so than with anything in the night sky. Heck, they *look* like lightsabers straight out of *Star Wars*. My green laser always stays in my inside jacket pocket when not in use (never loose, where someone can grab it), and the inevitable chorus of "can I try that?" is always answered with a polite "no."

THAT BEING SAID, GREEN LASERS AREN'T TOYS.

Never aim them at a person, an aircraft, or an automobile, as they can blind someone at close range.

FIRST LIGHT FOR A NEW TELESCOPE

I wouldn't get too ambitious with your new telescope on the first night out. Sometimes, just looking at the Moon (if it's up), whatever planets, and some star party faves like the Ring Nebula in Lyra or the Andromeda Galaxy are enough to get the feel for how your telescope works. I picked up my trusty 8-inch (20-cm) SCT from a photography store in Anchorage, Alaska. (This was before the era of online shopping, when you actually had to *drive* to the store to buy gear.) But it was a full week before I got my first views of the Ring Nebula from my driveway in North Pole, Alaska, through broken clouds. The joke and proposed axiom among amateur astronomers is that "the number of cloudy nights following the purchase of a new telescope is equal to its aperture in inches."

Assuming you've aligned your finderscope earlier in the day, you still might want to give it a double check using the first bright celestial object that pops out under the twilight sky. Things have a way of getting bumped in transport, and a slight tweak is easy enough to nudge things back into alignment.

Likewise, Newtonian telescope mirrors also need a quick tweak back into alignment, known as **collimation**. One caveat: Often, unless you're doing deep sky astrophotography, close is good enough for casual observing. Remember the physician's mantra of "first, do no harm . . ." I've seen many a telescope owner panic, wasting precious time under dark skies trying to get the collimation of their telescope mirrors just right. Often, this sort of obsession with perfect mirror alignment (and polar alignment) isn't really necessary for casual observing, and will only end up throwing things farther out of whack.

What to look at first? The very first objects that punch through the darkening twilight sky will be the Moon, planets, and brighter stars. We've even managed on occasion to start showing folks the Moon and brighter planets such as Venus *before* sunset. Venus also has less dazzle to it under a low-contrast twilight sky, another plus.

Collimating a Newtonian telescope: adjusting the set screws and using a laser collimator.

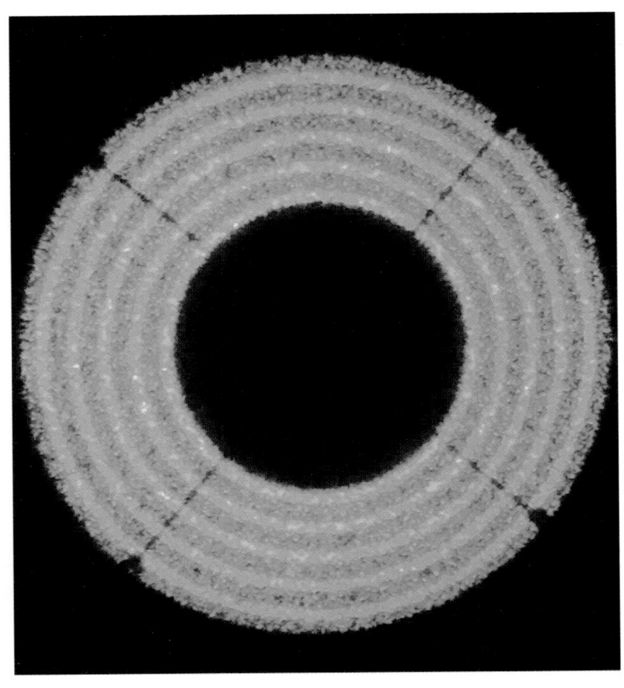

Star test: an out-of-focus view with well-aligned optics should look like concentric doughnut rings, with a well-centered hole.

Sometimes, the skies are only partly clear and we're chasing holes in clouds—also known as sucker holes. About one week a month, the Moon is near Full phase and can also hamper deep sky observations. Some hardcore astrophotographers simply pack it in during the week surrounding the Full Moon, which I think is a shame, as the Moon is interesting in its own right.

The very best time to observe the Moon, however, is near Quarter (half) phase, when the peaks and craters along the **terminator** (the line separating night from day) cast shadows and stand out in stark contrast. Through a telescope, the Moon looks like a real world right around this time, with jagged peaks and shadowed crater floors, begging for adventure. It's dramatic to watch the tip of a lunar mountain peak catching the first rays of sunrise, while its base is still shrouded in shadow. On the Moon, however, that Sun will continue to shine until sunset, two weeks later. Looking up, standing on that peak, you'd see the Earth in an opposite phase (i.e., a waning Last Quarter Earth versus a waxing First Quarter Moon) about 2 degrees across, four times as large as the Full Moon as seen from the Earth.

If you're showing off the Universe to a public group under the less-than-optimal conditions described above, I'd advise you to have a good repertoire of bright double stars handy. These are great talking points (remember the double sunset as seen from Tatooine in *Star Wars*?) and also serve to remind us that a majority of stars you see in the night sky are, in fact, double or multiple systems, and that our Sun is an oddity in the solitary minority. Plus, some multiple star systems are simply amazing to contemplate: Castor in Gemini, for example, is a sextuplet system, though you'll only see three at the eyepiece.

And speaking of eyepieces, one of the biggest favors I've ever done myself is upgrade my 8-inch (20-cm) SCT from its standard 24 mm eyepiece to a larger 42 mm model. This gives it a field of view as large as the Full Moon, making it much easier to sweep things up. There's nothing more frustrating than trying to perform extraordinary acts of contortion in the dark while trying to look through a pinhole eyepiece.

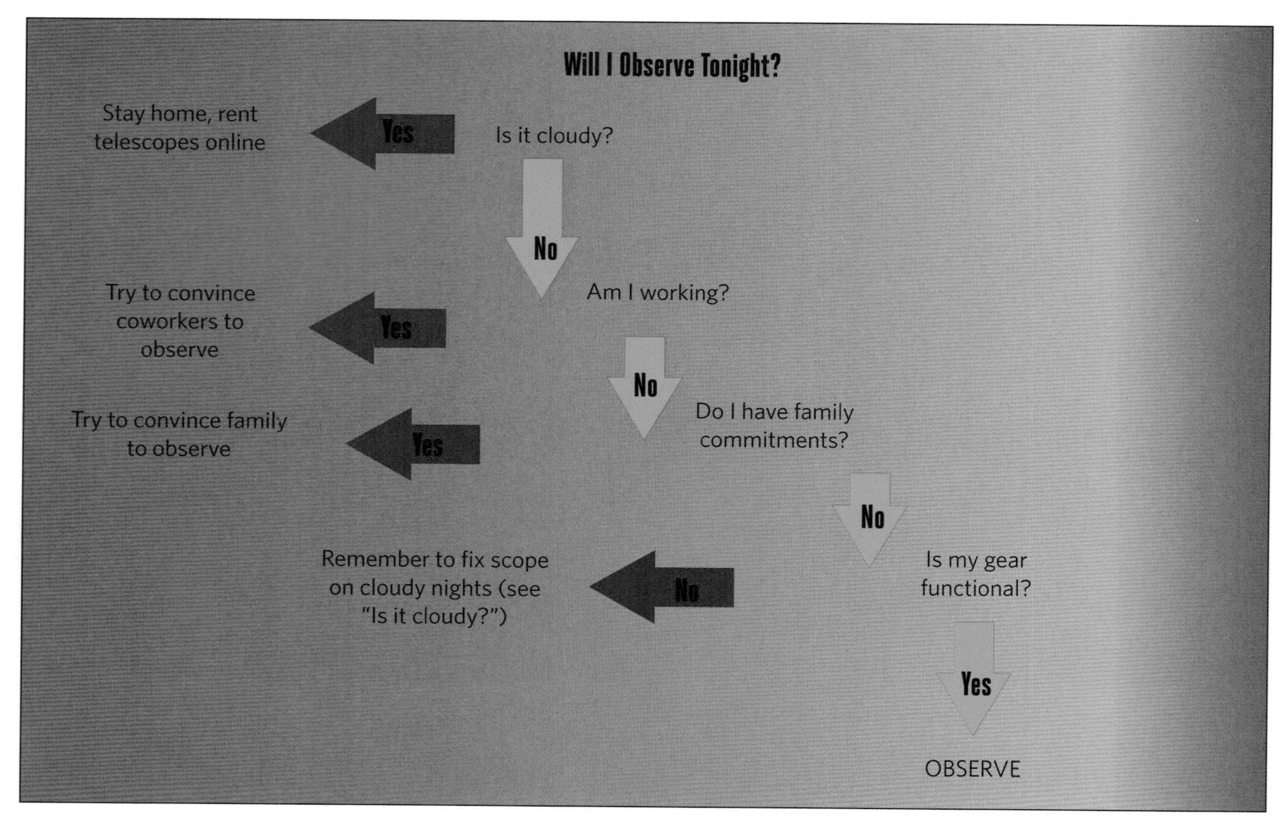

Will I observe tonight? A go/no-go logic tree.

A stepladder is also a handy accessory in the field, or in the case of a giant full-size Dobsonian telescope, a full-size ladder. Again, we like how the SCT fork-mounted configuration puts the eyepiece down to a reasonable compromise in terms of viewing height, no matter where it's aimed in the sky. (Probably its biggest blind spot is around the zenith to the celestial pole.)

Anticipation also plays a big part of knowing what to look for at the eyepiece. Think of a hunter stalking prey, a well-camouflaged adversary that often doesn't want to be found. Globulars look like powdered sugar sprinkled on the sky; a nebula often looks like a fuzzy smudge, or in the case of a planetary nebula such as Messier 57, the ghost of a doughnut. An evenly matched close double-star pair looks like a distant set of car headlights on a faraway hilltop. (These are all real metaphors I've heard at public star parties. Kids are often the best when it comes to descriptive observation.) Knowing what you're looking for and what it *should* look like can go a long way toward actually finding it.

On the flip side of the coin, also keep an eye out for the unexpected. Alan Hale and Thomas Bopp, for example, just happened to be casually observing the globular cluster Messier 70 one evening on July 22–23, 1995,[13] when they both independently stumbled across a fuzzy smudge that stubbornly refused to snap into focus. That smudge went on to become one of the greatest comets of the twentieth century, Comet Hale-Bopp. Anthony Wesley noticed a mark of Jupiter that hadn't been there the night before back in 2009,[14] and alerted astronomers worldwide of an impact on Jupiter. We even stared right at a supernova in the galaxy M82 during *Universe Today*'s Virtual Star Party . . . while broadcasting it live![15] (We're still kicking ourselves for the "one that got away . . .")

Finally, we want to mention one of the backyard astronomer's greatest nemeses: dew. If you're lucky enough to live out in the dry desert of the Southwest, we'll describe the foreign concept of dew as droplets of moisture that collect on surfaces (including precious telescope optics) at night. If you live on the damp U.S. Eastern Seaboard, then dew is a familiar nightly reality which often collects fastest on those crystal-clear nights that are just beckoning for you to set up your telescope and observe. The reason for this is because a clear night is often also a cool one, as the heat of the day radiates away faster when it's unimpeded by a blanket of clouds. The ambient temperature drops below the dew point, and drops of water condense out of the moisture-laden air.

In the case of a car parked out in a Florida driveway overnight, this problem is easy to solve: you wipe it off, roll your windows down, maybe run your defoggers a bit to warm the glass up over that dew point, and you are on your way.

Keeping telescopes fog- and dew-free is trickier. You never want to just wipe them off, as small particles of dust and dirt can scratch optics. I once stood aghast as a well-meaning friend tried to scrape frost off of the corrector plate of my telescope, shortly after bringing it in out of the frigid Alaskan night.

The goal when it comes to dealing with dew on a telescope is simple—you want to insulate and preserve those cooling optics for as long as possible, while keeping them just above the dew point. Several electric dew-control systems are available that position heating strips around optics and eyepieces. If I'm near a power source, I like to use a hairdryer to remove dew about once every half hour or so. Make sure, however, that you use the lowest setting, and aim the hairdryer at the optics at an oblique side angle. Go at it with high-powered gusto, and you could crack or distort your optics from overheating them too quickly!

I also like to use a foam hiking mat to insulate the telescope's tube out in the field. I use either Velcro or bungee cords to keep the cover firmly in place, but be careful with bungee cords, as setting them too tight can cause them to snap back quickly (and smack unsuspecting operators or bystanders with a nasty surprise) on release. One-inch-thick foam hiking mats cost only a few dollars, can be easily cut to size with a box cutter, and will do the same job as a $30 telescope dew shield.

The Trifid Nebula, Messier 20. This 40-minute exposure not only brings out the delicate colors in the emission/reflection nebula but the contrasting dusty, dark lanes as well.

All of this moves toward the final goal of optimizing your time out under the night sky. Time is precious, and we never want to let a clear night slip by. Dedicated amateur astronomers continually run a sort of mental Drake equation in their heads when it comes to observing, factoring in the fraction of nights that clouds, work, family commitments, and gear breakdowns keep us from observing. The bright phase of the Full Moon and access to a good dark sky site also plays a role, though many urban and suburban sites are now so light-polluted that the addition of the Moon doesn't really matter.

We can optimize that last factor when it comes to equipment by making sure our gear is ready to go.

ACTIVITY: BUILDING AN OBSERVATORY FOR UNDER $500

Are you a weekend warrior when it comes to home improvement? I spent 20 years enlisted in the U.S. Air Force working on weapons systems. I therefore have a Tim Allen sort of approach when it comes to handiwork around the house: If you want something to explode when it's supposed to, then I'm your guy. Maybe it's better to hire a contractor, though, to build a new bathroom.

But when it comes to backyard observatories, they need not be elegant or expensive, merely functional. The key asset an observatory gives you is convenience: Simply unlock it, slide off the roof, and turn the scope on, and you're observing. When you're done, the telescope stays aligned, set up, and protected from the elements, ready to go for the next session.

Most observatories come in two types, with either a dome or a roll-off roof. A third hybrid version is where the telescope is stationary, and the shed itself slides back, exposing the telescope to the sky. A dome is iconic; it simply *looks* like what people expect an astronomical observatory to look like. The dome rotates and a slit opens, as the telescope follows the view.

A slide-off roof, however, is simpler, and exposes your telescope to the entire sky. While a closed-in dome makes sense for a large professional observatory located on a windswept mountaintop, a slide-off roof is far easier to construct, as you don't have to synchronize the telescope to rotate in tandem with the dome.

Step 1: Gathering the Hardware

I built our aptly named Very Small Optical Observatory (VSOO) out of a modified 9-foot by 10-foot (2.7-m by 3-m) rectangular Sears garden shed. A prefabricated vinyl shed would work as well, as long as you can make the key modification for the roof to slide off. This was a simple two-day, one-person project, and the cost covered the shed, lumber, locks, and hardware. I had the local hardware store cut the floor sections out of plywood as per the instructions with the shed kit. The key modification (where I like to say "we're now leaving the instructions for parts unknown") was to make shed roof slide off. To this end, I assembled the roof separately and used a set of rail brackets to reinforce it, as the roof was a bit flimsy on its own.

Supplies and Tools Needed

- One shed kit
- Plywood flooring
- Tape measure
- Hand saw
- Drill
- Phillips screwdriver
- Hammer and nails
- Saw horses
- Set of two to five locks, all set for the same key
- 2x4 and 4x4 lumber (Cut the 2x4s into five 9-foot (2.7-m)-long pieces for the tracks for the observatory roof to slide on. The 4x4s create the clothesline brace for the roof to sit on while the observatory roof is open.)
- PhD (post hole digger)

Step 2: Construction

I assembled a clothesline truss to the side and back of the observatory using posts and 2x4s. This is what the observatory roof sits on when it's slid back and open to the sky. Though the observatory is aligned with its longest end north-to-south, the roof slides off to the east, so the peak doesn't block polar alignment. Four added padlocks keep the roof locked and closed when not in use, and bumpers on the end of the clothesline truss keep the user from sliding it back too far in the dark.

The Very Small Optical Observatory (VSOO) in its heyday outside of Vail, Arizona. This sequence shows key stages of construction, from (clockwise) building the floor (upper left) to mounting the slide-off roof, to the finished observatory with the roof closed and the roof open, exposing the telescope to the Arizona night sky.

Step 3: Telescope Installation and Operation

This setup worked admirably for years, until we sold our house and moved from Arizona. Our 10-inch (25-cm) Schmidt-Newtonian telescope was a permanent fixture in the observatory, and I could have it up and running in 5 minutes. I simply had the scope set on its tripod, with floor-mounted metal brackets screwed into the plywood floor to dampen the points of the tripod legs and keep them fixed in position. You would get some slight vibration with this configuration as people (and our four dogs) walked around inside the observatory, but it was simple and effective. For astrophotography, I simply stood outside of the observatory and triggered the camera release using a remote cable.

Step 4: Possible Modifications

We planned to ultimately build a ground-mounted concrete pier for the telescope separated from the floor, before selling the house. It would have been an easy add-on to install. We also had to be careful when we removed the roof in high winds, as it simply sat on the rails. Another easy modification would have been to add a garage door opener to open and close the observatory.

Still, even such a simple observatory increased my observing time exponentially, which was a great plus, and what having your own observatory is really all about.

FOLLOWING PLANETS AND THE MOON IN THE SKY

What to watch for (and how to find) key features on the Moon and planets in our Solar System, plus observing eclipses, occultations, and more.

"I have made it a rule never to employ a larger telescope when a smaller will answer the purpose."

— Sir William Herschel, discoverer of the planet Uranus

They're some of the first objects that most amateur astronomers aim a telescope at. The Moon and the planets of the Solar System are old friends, moving through the thirteen astronomical constellations along the plane of the ecliptic. The positions of the Moon and planets change from night to night, season to season, and year to year. Chances are there are at least one or more planets visible in the night sky *tonight*, as they traverse between the dawn and dusk sky, and back again.

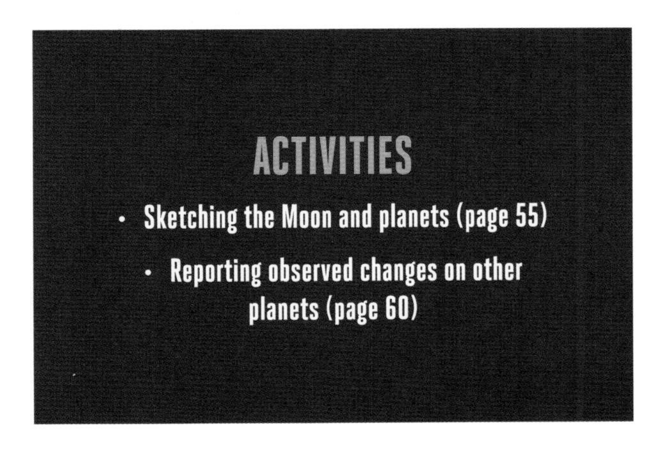

THE MOTION OF THE PLANETS

Ancient astronomers noticed that several stars in the sky refused to stay put. The Greeks gave these vagabond luminaries the name *planetes*[1] (literally "wanderer"), leading to our modern-day names for the planets. Five planets are visible to the naked eye: Mercury, Venus, Mars, Jupiter, and Saturn. The advent of the telescope added the ice giants Uranus, Neptune, and—from the span of time from its discovery in 1930 to its 2006 demotion—Pluto to the Solar System family of worlds known as planets. The asteroids 1 Ceres, 2 Pallas, and 4 Vesta were briefly considered planets shortly after their discoveries in the nineteenth century. Of course, so was the spurious, hypothetical world of Vulcan supposedly orbiting between the planet Mercury and the Sun. The definition of a planet has been traditionally more of a cultural than a scientific one, as highlighted by the recent controversy surrounding Pluto. The Solar System has eight major planets . . . for now.

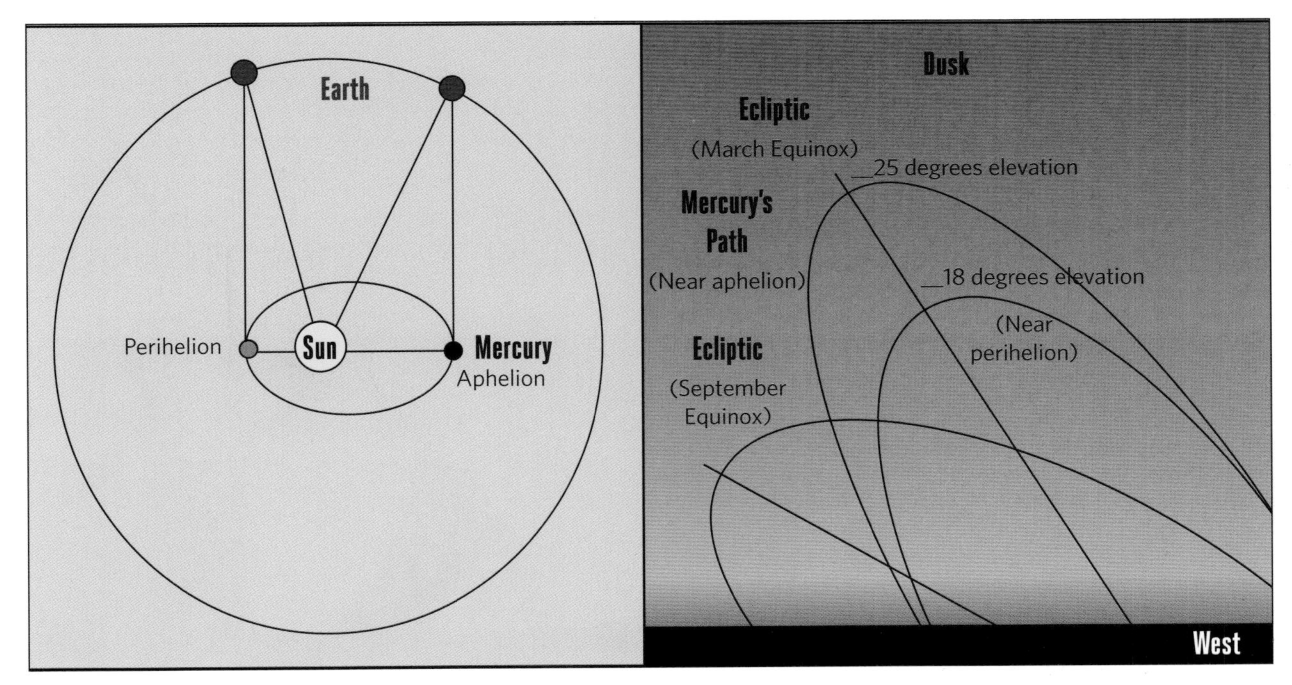

Not all apparitions of Mercury (and Venus) are favorable. If a planet reaches greatest elongation near aphelion (shown in the left panel), for example, it appears farthest from the Sun. The seasonal angle of the ecliptic plane versus the horizon (right panel) also plays a key factor in whether the planet is "in the weeds" at dusk or dawn, or up out of the murk low to the horizon. Mercury is especially prone to this effect.

The interior planets of Venus and Mercury never stray far from the Sun, so they're only visible in the dawn or dusk sky. Mercury is the most elusive of the naked eye planets—one (probably apocryphal) tale[2] says that fifteenth-centry astronomer Nicolaus Copernicus never saw Mercury for himself. Inner planets reach what's known as **greatest elongation** east (dusk) or west (dawn) at their greatest apparent extent from the Sun, an ideal time for observing them. Think of a right triangle, with Mercury or Venus at the apex of its two legs, and the Earth and the Sun at either end of its hypotenuse. Though difficult, there are a few favorable elongations of Mercury every year, ideal times to check the innermost world off of your life list if you know exactly when and where to look for it low in the dawn or dusk sky.

The innermost planets can also reach **inferior conjunction** between the Earth and the Sun, and **superior conjunction** on the far side of the Sun. The orbits of Mercury and Venus are tilted slightly with respect to the Earth, but occasionally, they can cross the visible disk of the Sun in what's known as a **transit**. Transits of Mercury[3] are much more frequent than those of Venus: Mercury transits the Sun thirteen times in the twenty-first century, while Venus last transited the Sun on June 5–6, 2012, and won't do so again until 2117.

Outer planets move at a much more leisurely pace through the night sky. The very best time to observe an outer planet is when it nears **opposition**, a point at which it lies closest to the Earth and rises in the east, as the Sun sets opposite in the west. Oppositions for outer planets such as Saturn and Jupiter occur nearly every year as the speedy Earth laps these worlds once per orbit, while oppositions of Mars occur roughly once every 26 months. When planets reach a point 90 degrees east or west of the Sun, we say they're at **quadrature**. Think of the same right triangle again, this time with the Earth at the apex and the Sun at the end of one vertex, and the given planet at the end of the other.

Outer planets always present their illuminated face toward the inner Solar System and the Earth. You'll never, for example, see a crescent Jupiter or Saturn from the Earth, though Mars is close enough to us that it may look slightly gibbous near quadrature. Venus and Mercury, however, can show phases like the Moon, as we see the nighttime sides of these inner planets as they pass between the Earth and Sun.

The roll call of "what is a planet?" has changed over the centuries as well, as we've come to realize our true place in the Solar System. In the geocentric scheme of the Universe laid out by the ancient Greeks, Earth stood at the unmoving center, with the Sun and Moon relegated to the role of planets along with Mercury, Venus, Mars, Jupiter, and Saturn. The advent of Copernican heliocentrism in the early sixteenth century swapped roles for the Earth and Sun: The Moon was now a satellite of the Earth (correct), which was just another planet (correct again), going around the Sun at the center of the Universe (not quite true, but close enough when you're describing the local solar neighborhood). Remember, medieval astronomers in Copernicus's time wouldn't even have the telescope available to them for another century. It's remarkable that they understood as much as they did, using only naked eye observation.

Copernican heliocentrism was a simpler, more elegant way to describe what we actually see going on overhead in the sky . . . to a point. It was certainly easier than ascribing the motions of the planets as sitting in a convoluted schema of interlocking crystalline spheres (some systems needed twenty-eight spheres or **epicycles** per planet to make this system work!) going around Earth, perched atop a tortoise, perched on a camel, perched on an elephant, perched on a . . . well, you get the idea.

A flat-Earth proponent once replied to philosopher William James that it's simply "turtles all the way down . . ."

The glorious planet Saturn, rings and all. Shot through a 12" (30-cm) LX200r telescope with 2.5x Powermate and Chameleon 3 camera with LRGB filters.

A drawing of William Herschel's 40-foot (12-m) telescope (left), along with the original 48" (1-m) diameter speculum mirror, now on display in the Science Museum in London, England, United Kingdom (right).

At last: our best views of Pluto (left), its large moon Charon (upper right), and the tiny moons (inset) Nix (left) and Hydra (right) from the July 2015 flyby carried out by NASA's New Horizons spacecraft. (Not pictured in this family portrait of the Pluto system: Styx and Kerberos.)

Still, even using a heliocentric model, the pesky planets refused to stay put. This anomalous motion would finally be solved by Johannes Kepler, who proposed that planets move in ellipses around the Sun, moving faster closer in to the Sun (near perihelion) and slower near aphelion farther away.

Kepler didn't like his own solution. Certainly, circles are tidier, and reflect the perfection thought adherent in the heavens. The trouble was, those annoying ellipses *worked*. This revelation, more than anything else, was the first step toward giving the science of astronomy the ability to backup observation. Kepler was the last astrologer, and also the first astronomer.

But the roll call of the planets has still been a revolving door, right up to the current day. William Herschel added the first planet discovered by a telescope to the Solar System on March 13, 1781, when he spotted Uranus. We can be thankful that his proposal to name the planet *Georgium Sidus* (George's Star), after his benefactor King George III, didn't stick.[4]

On the first day of the nineteenth century (January 1, 1801), the astronomer Giuseppe Piazzi discovered an +8th magnitude "star" moving through the constellation Taurus from night to night, an object that became known as 1 Ceres. The ranks of these faint wandering stars continued to grow, and soon included 3 Juno, 4 Vesta, and 2 Pallas, all considered planets in the early nineteenth century.

Then, the French theorist Urbain Le Verrier scored one for France, when he predicted that the anomalous motion of Uranus could be explained by the gravitational pull of an as-yet-unseen body. "Look here, at this patch of the sky on this date," was the gist of Le Verrier's prediction, "and you'll see a new planet."

The astronomers Johann Galle and Heinrich d'Arrest took Le Verrier up on the challenge, and spied Neptune from the Berlin observatory on the night of September 23–24, 1846. This blew the figurative barn doors off of the astronomical community. Newtonian physics now gave astronomy true predictive power. Le Verrier, having achieved rock star status, went on to predict the inter-Mercurial world of Vulcan for an encore. This spurious world was thought to be tugging on Mercury's orbit, the story goes, hiding close to the Sun.

Though Vulcan proved to be anomalous and these tiny nineteenth-century worldlets were eventually reclassified as **asteroids**, they were still often listed as planets by many astronomy textbooks of the day. Pluto—another tiny newcomer on the Solar System scene—was discovered by the astronomer Clyde Tombaugh in 1930 and shared the same fate, enjoying planetary status for several decades before demotion to the rank of dwarf planet by the International Astronomical Union (IAU) on August 24, 2006.[5]

Today, we know that the orbit of Mercury can be described by Einstein's theory of general relativity, which explains how the mass of the Sun warps the fabric of space.

That's a 60-second tour of the history of astronomy and the Solar System . . . now, how will *you* know how to find those planets tonight?

Of course, the Moon is easy to identify, no GoTo mounted telescope is required for that one. How about finding the planets? Here, knowledge of the static patterns of the constellations is key. Remember the handy rule that "stars twinkle, planets don't." Looking at, say, the familiar shape of the constellation Leo the Lion, it might be easy to spot a bright interloper such as Jupiter rivaling the star Regulus for its kingship.

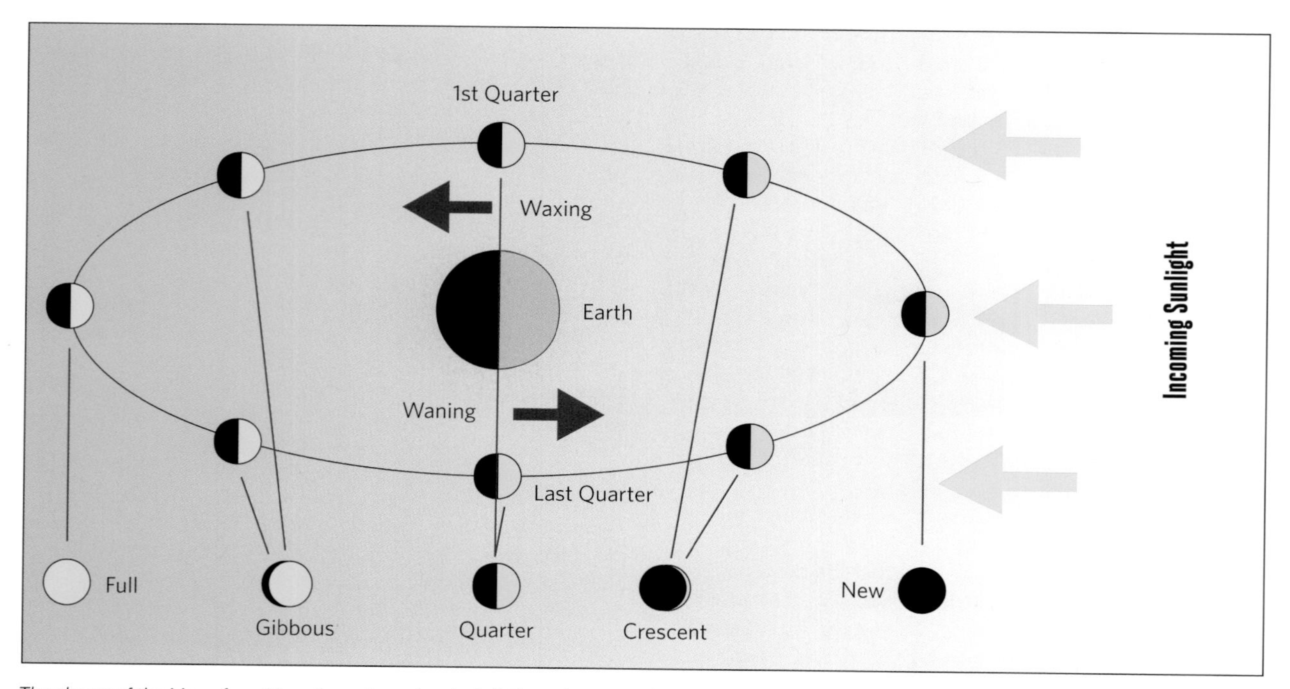

The phases of the Moon from New, through waxing, to Full, through waning, back to New, over the span of one synodic month (29.5 days). The size of the Moon's orbit is not to scale.

Planets have their own unique visual attributes as well. Venus, for example, is brighter than any natural object in the sky, other than the Moon and the Sun. Both Venus and Mercury never stray far from the Sun, and are always found in the dusk or dawn sky. In the case of a fleeting world like Mercury, you can see it move rapidly against the starry background from one night to the next. Mars is noticeably ruddy in color, and only comes around to a favorable view once every 26 months. Giant Jupiter is brighter than any of the stars along the ecliptic, and moves with a stately motion. Saturn is fainter and slower still, with a yellow-saffron appearance.

In 2011, I noticed Saturn approaching the bright star Spica in the constellation Virgo, and realized I had followed it for one full 29-year orbit, bringing it back full circle to the same spot I started following it as a teenager waaaaaay back in 1982. (That dates me, I know!)

Here's a world-by-world rundown of highlights to watch for at the eyepiece:

THE MOON

Through binoculars or a small telescope, the surface of the Moon leaps out at you, waiting to be explored. Even a small telescope readily reveals the Moon as a *real world*, a place that looks like it's close enough to reach out and touch. During the late 1960s and early 1970s, twelve people did just that, landing on the lunar nearside and returning samples to the Earth during NASA's Apollo program.

INCONSTANT MOON

Another change that is readily apparent in the sky is the motion of Earth's moon from night to night. The Moon's orbit is inclined about 5 degrees relative to the ecliptic, meaning that the Moon can actually appear in eighteen constellations (those extra six non-zodiacal constellations are Orion, Auriga, Sextans, Corvus, Crater, and Ophiuchus) in addition to the twelve traditional constellations of the zodiac. The motion of the Moon is simple at a glance, then wonderfully complex upon close examination, giving early astronomers a great lesson in Celestial Mechanics 101.

The most basic observable phenomenon is the **Moon's cycle of phases** as it orbits around the Earth once every **synodic month**, a period of 29.5 days. The Moon's phase is simply the portion currently illuminated by the Sun that is turned Earthward. The Moon is New when its ecliptic longitude equals that of the Sun, with its nighttime hemisphere turned Earthward, and the Moon is Full when it's near 180 degrees opposite from the Sun. When the nodes, where the path of the Moon meets the ecliptic, fall near these points, a syzygy allowing for an **eclipse season** occurs. Solar eclipses always occur on a New Moon, and a lunar eclipse always occurs at Full Moon. When the Moon is fattening up through the early evening sky, we say it's **waxing**, while a thinning Moon in the early morning sky is referred to as **waning**. A half Moon is also either at First Quarter (waxing) or Last Quarter (waning). Think of the quarter as one fourth of the Moon's total synodic cycle. The fingernail-shaped Moon between New and First or Last Quarter is referred to as the crescent Moon, while the football-shaped Moon before or after Full phase is a waxing or waning gibbous Moon. Keep in mind that the points of New, Full, and First/Last Quarter are mere transitional moments in time, while the Moon spends the entirety of its orbit either waxing or waning, gibbous or crescent.

Does the Moon look larger to you when it's rising or setting? This is the famous **Moon illusion**, a variation of the Ponzo illusion where the human eye and brain judges the size of an object (in this case, the rising Full Moon) based on the foreground. You can easily refute this by shooting a sequence of images of the rising Moon in which the Moon illusion completely vanishes. You can also hold out your hand at arm's length and see how your pinky fingernail just covers the Moon. Try this when the Moon looks "huuuge," and then try again when the Moon is much higher and doesn't seem so big. It's the same size.

Fun fact: The Moon is actually *closer* to the observer by about one Earth radii (~4,000 miles [6,400 km]) when it's overhead near the zenith, versus when it's on the horizon.

DOES THE MOON LOOK LARGER TO YOU WHEN IT'S RISING OR SETTING?

A fine view, filled up with a waxing gibbous Moon. Shot through a Celestron Edge HD11 telescope using a Canon EOS 6D camera. Two-panel mosaic processed in Photoshop.

FINDING YOUR WAY AROUND THE LUNAR LANDSCAPE

The shadow play of the Sun's illumination changes through the 29.5-day **synodic period** of the Moon, as it works its way waxing from New to Full, then waning back to New again. For the purposes of this guide, we'll refer to "east" and "west" on the Moon from the observer's perspective, against the sky. The very best time to look at the Moon is near First or Last Quarter (half) phase, when mountains and crater rims along the terminator—the line dividing night and day—stand out in stark contrast. Still, it's worth checking out the Moon at every phase—even near Full—if it's in view, as the lighting angle makes alternating features seem to appear, then vanish from view from one night to the next.

The Moon's orbit isn't perfectly circular. Like all celestial objects, it reaches perigee (its nearest point) and apogee (its farthest point) once per orbit. The nearest Full Moon to perihelion for a given year has become known in recent years as a **Supermoon**, and the most distant Full Moon of the year is also referred to as a **Minimoon**. Although it generates lots of interest every year, the Moon only varies from 29.3 to 34.1 arcminutes in apparent size.

The Moon may seem a pearly bone-white and almost painfully bright at the eyepiece near Full phase, but this illusion is the result of having sunlight reflecting back at us, concentrated from a relatively small source. The pin light setting in Adobe Photoshop mimics a similar effect. The actual surface reflectivity of the Moon (known as **albedo**) is about 14 percent, similar to that of worn asphalt. In fact, Apollo astronauts described the texture and color of lunar dirt as similar to that of coal dust.

Looking at the overall face of the Moon, the first features that jump out at the observer are the relatively dark **maria**, contrasted with the bright **lunar highlands**. The Earthward face of the Moon is distinctly different than its almost maria-free **farside**, as revealed by Russia's Luna-3 when it flew past the Moon on October 7, 1959.

A flurry of impact craters also immediately jump at you during even a casual glance at the Moon at low power. Rare on Earth due to erosion, the Moon preserves an ancient record of these impacts that have pummeled the inner Solar System for billions of years. Earth gets hit with the same frequency as well, but glaciers, wind, and rain all serve to cover these up after just a few thousand years.

"Maria" comes from the Latin term for seas, which is what Galileo thought the smooth surfaces were. They certainly do have romantic names to them, including monikers such as the Sea of Crises and the Sea of Storms. Other ominous names such as the Marsh of Epidemics (*Palus Epidemiarum*) and the Lake of Hatred (*Lacus Odii*) dot the surface of the Moon, perhaps to vex any future lunar colonists that call such obscure locales home.

The Greeks held that the face of the Moon was somehow the reflection of the Earth below—this would be tough to explain though, just how the Earthward face of the Moon looks the same from every angle. What *is* a reflection of the Earth, however, is a phenomenon known as **earthshine** or **Ashen Light** seen on the night side of the Moon, during crescent phase. This is sometimes also referred to as the "Old Moon in the New Moon's Arms."

I always thought it was curious that no one ever attempted to produce a realistic face of the Moon in pre-telescopic times, though they would certainly try later in the case of other planets in the post-telescopic era, albeit with a smaller and murkier view.

Newer craters (we're talking craters only a few million years old) overlap older ones, and serve as a great tool to date terrain on the Moon. Some older craters overlap and have collapsed central peaks, or are missing rims entirely. Scanning the lunar surface at high magnification, you might even come across **crater doublets** or **crater chains**, commemorating a string of ancient impacts on the lunar surface.

HUNTING FOR THE (MEN) ON THE MOON

It's a question I get nearly every time I show off the Moon to the public. "Can you see the U.S. flag the astronauts put up there with that thing?" The answer, unfortunately, is no, though maybe someday, future lunar colonists will arrange a gigantic flag of the lunar surface just for Earthbound star parties. They're simply too tiny. The base of the lunar lander—the largest piece of hardware left on the Moon—is only 9.4 meters across, far smaller than the resolution of even the largest Earthbound instrument. Remember our friend the Dawes Limit? NASA's Lunar Reconnaissance Orbiter has managed to image the lunar landing sites from its vantage point in low lunar orbit. It is, however, fun to scout out the same terrain that the Apollo astronauts once crossed, and at high power, you might just spy the trio of 5-kilometer craters named after the Apollo 11 astronauts Armstrong, Aldrin, and Collins adorning the Sea of Tranquility.

It's often said that the Moon keeps one face turned toward the Earth, though this is also only approximately true. The Moon also rocks slowly back and forth as it orbits the Earth, in a process known as **libration**, and nods side-to-side like a slow-motion headbanger in a motion known as **nutation**. This assures that we actually see 59 percent of the surface of the Moon as we peer over its sides, though we always see the craters at the poles and along the edges at an oblique angle.

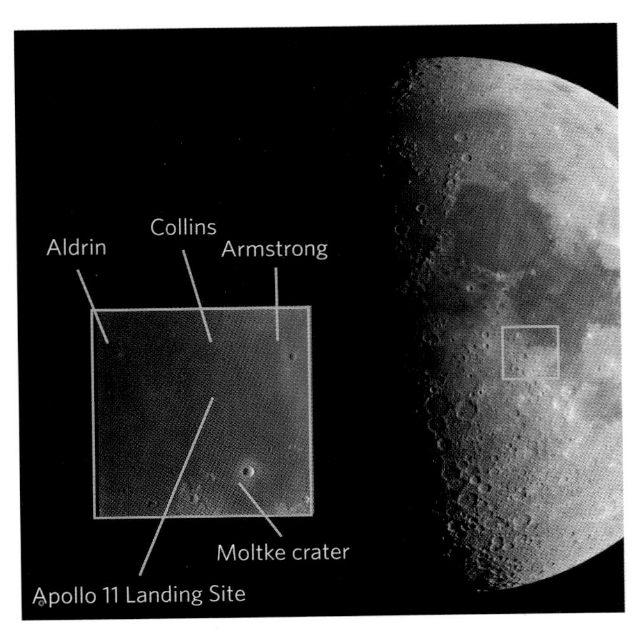

Finding the craters of Apollo, near the Apollo 11 landing site.

To date, all crewed missions to the Moon have landed on its nearside. We've still got lots of the Moon left to explore.

Here are the top ten lunar features to check out. A great deeper dive for serious lunar explorers is the U.S. Geological Survey's Astrogeology Science Center (https://astrogeology.usgs.gov/maps), a fine place to brush up on **selenography** (the study of the Moon), as well as the surfaces of other planets in the solar system.

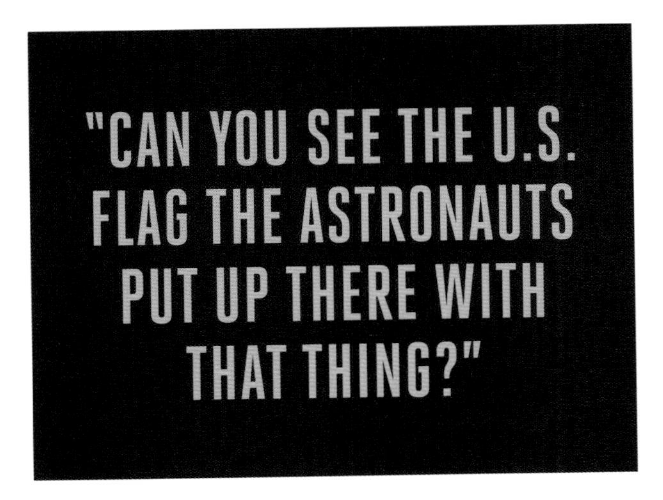

10. EARTHSHINE

This is the reflection of sunlight off of the Earth, seen on the nighttime side of the Moon. You'll see lots of confusion between the term (cue Pink Floyd) dark side of the Moon, when people really mean the farside that's perpetually hidden from view as seen from the Earth. We can indeed see the dark night side of the Moon, as it's merely the portion not currently illuminated directly by the Sun. The portion shaded gray by earthshine can vary in brightness, due to the amount of brilliant snow and cloud cover on the Earth currently turned moonward. Best view: crescent phase, 3 to 4 days before or after New Moon.

9. MARIA VERSUS HIGHLANDS

Don't overlook one of the most conspicuous sights on the Moon, one you can see with the naked eye. It's worth noting the overall contrast between the relatively dark lunar plains (maria), and the brighter highlands of the southern hemisphere. Best view: most every phase, though it's especially worth checking out this feature near Full.

8. MARE IMBRIUM

This maria in the northeast quadrant of the Moon makes up one eye of the Man in the Moon. Also keep an eye out (bad pun intended) for the dramatic curved scarp of the Bay of Rainbows (Sinus Iridum) tucked away in one corner. This is the filled-in basin of a giant impact crater. Mare Imbrium is also the resting place for China's Yutu (Jade Rabbit) rover, which landed on the Moon on December 14, 2013. Best view: when the Moon is waxing gibbous around 12 days old.

7. COPERNICUS

This is a prominent crater located in the southern edge of the Mare Imbrium region, a brilliant, relatively new impact festooned with brilliant rays. Copernicus crater is about 800 million years old. Best view: when the Moon is 11 days old.

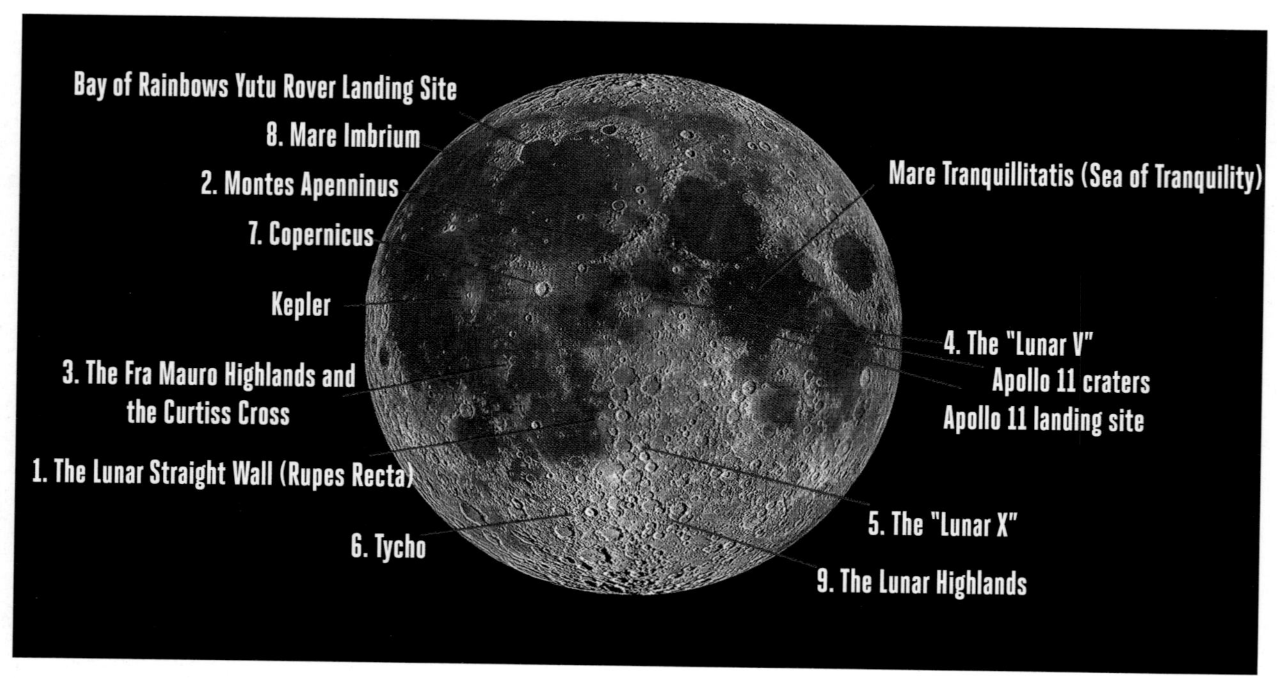

A map of the nearside of the Moon, with labels denoting key features.

6. TYCHO

Another notable crater in the lunar highlands southern region, prominent even at low power. Best view: when the waxing gibbous Moon reaches 10 days old.

5. THE "LUNAR X"

Sometimes also referred to as the Purbach Cross, the Lunar X feature is the result of the confluence of several ancient crater rims. It's a dramatic feature to catch as the first rays of sunlight hit it while its base is still shrouded in darkness. Best view: near First Quarter phase, when the Moon is 7 days old, about 6 hours prior to First Quarter phase. The precise time for First Quarter phase for each lunation is generally listed on a given Moon phase table, sometimes in local time, sometimes in Universal or Greenwich Mean Time (UT/GMT).

4. THE "LUNAR V"

Another lunar letter feature, this V-shaped prominence located on the edge of the Mare Vaporum often stands out right along with the aforementioned Lunar X. Best view: near First Quarter phase, when the Moon is 7 to 8 days old.

3. THE CURTISS CROSS

An elusive feature adorning the Fra Mauro highlands near the landing site of Apollo 14. Near Parry crater, also keep an eye out for a unique crater doublet very near where X marks the spot for this old collapsed mountain chain. Best views: right around when the Moon is 11 days old.

2. MONTES APENNINUS

The Lunar Apennines are a dramatic mountain chain at the edge of the Mare Imbrium. Just imagine hiking along these, with the crescent Earth high overhead. Best views: these can be dramatic right around early waxing gibbous phase, when the Moon is 9 days old.

1. THE LUNAR STRAIGHT WALL (RUPES RECTA)

One of our favorites, this sword-shaped feature gracing the Mare Nubium stands out when the Sun hits it just right. What you're actually seeing is a 72-mile (116-km) ridge about 1,200 feet (400 m) high. Best view: when the Moon is 8 days old.

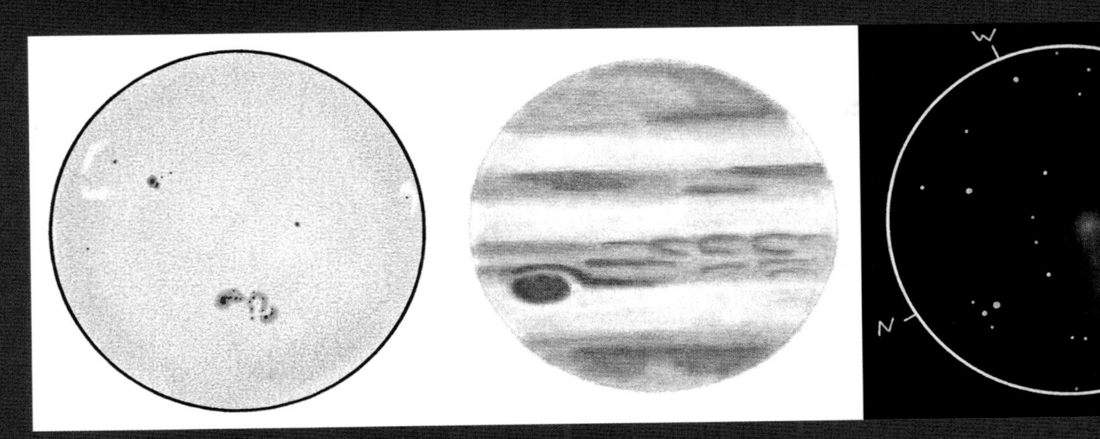

Examples of three sketches of three different types of astronomical objects:

1) The Sun on August 7, 2015. Done with a NexStar 6SE Schmidt-Cassegrain telescope using a 25 mm eyepiece. The large sunspot group located at the lower centre of the Sun's disc is active region 2396, which was about three times the size of the Earth at the time.

2) Jupiter on May 1, 2016, showing the Great Red Spot to the lower left. Done using a NexStar 6SE Schmidt-Cassegrain telescope with a 6 mm eypiece.

3) Comet C-2014 Q2 Lovejoy passing through the constellation of Triangulum on January 26, 2015. Done through a AstroMaster 130 mm reflector telescope using a 32 mm eyepiece.

ACTIVITY: SKETCHING THE MOON AND PLANETS

Want a permanent record of what you see at the eyepiece? Consider approaching the Moon and planets just like astronomers of yore did, and try your hand at simply sketching what you see. Astronomy is one of the few sciences that lends itself to artistic expression. All of the early astronomers were dabblers in the sense that they tinkered as inventors, alchemists, and naturalists as well, and that meant drawing what they observed in the world around them.

Whether you're an astrophotographer, sketcher, or simply keeping a log or blog about your astronomical exploits, you're carrying on in that great tradition of showing others what the Universe around you looks like from your special vantage point tonight. A feat of visual athletics, like, say, the astrophotographer Thierry Legault carefully planning to catch a transit of the International Space Station across the partially eclipsed Sun from the deserts of Oman,[6] may not have much in terms of scientific value, but we can still marvel at the execution and the fact that such bizarre sights are indeed happening around us all the time, if we know just where and when to look for them.

Why draw what we see at the eyepiece? Can't modern DSLRs top what the hand and eye can produce?

Especially in terms of the Moon and the planets, sketching still has its place:

- Sketching a subject trains the eye to see detail you might otherwise pass up at a glance.
- Sketching is a neat way to see the movement of a comet or an asteroid: Simply sketch the suspect star field on one night, and then the next. Check whether any of the stars moved against the background.
- Sketching is a great record to go back to, even years later.
- In terms of planets or the surface of the Moon, your eye can still sometimes tease out detail that even advanced image processing programs might miss.
- Unlike astrophotography, there's a very low cost buy-in to astronomical drawing.

I'll admit, I'm no artist. I've seen some sketches done at the eyepiece that look like actual photographs; ours are still very far from that level of achievement. And while deep sky sketches are often best with a white-on-black paper background, you can pull off planetary sketches using a plain old No. 2 pencil and paper, with maybe a blending stump to mottle the edges a bit.

Some pre-drawn circles are handy as well, as these serve for templates for either the field of view at the eyepiece or the planet itself. Mars near opposition, Jupiter, and Saturn all lend themselves to changing details that are challenging to draw. In addition to a tablet, pencil (pen if you're brave, but remember, you'll probably be erasing lots), and colored pencils, you'll need a small table in the field. A headlamp is also necessary, and unlike with deep sky observing, you can use a white light while drawing the Moon and planets, as light pollution isn't really much of an issue when you're observing bright sources.

Finally, don't be afraid to digitally scan and share those artistic masterpieces of the sky with the world out across social media.

Happy sketching!

A PLANET-BY-PLANET RUNDOWN

The Planets: The moons of Jupiter, phases on Venus, even stubby hints of Saturn's rings are all within the grasp of binoculars or a small telescope. As we've mentioned, the planets wander the plane of the ecliptic, appearing in the twelve traditional constellations of the zodiac, plus Ophiuchus. We say that the apparent motion of a planet is **direct (prograde)** when its moving eastward, and a planet is in **retrograde** when it's moving in a reverse motion westward.[7] Astrology-minded folks make a big deal about, say, "Mercury in retrograde," but this motion is only an illusion based on our Earthly perspective, as our planet overtakes (or is overtaken by) a passing world, like runners lapping one another on a racetrack.

Rather than blaming the heavens on our Earthly woes, it's better to take a hint from Shakespeare's *Julius Caesar*: "The fault, dear Brutus, is not in our stars, but in ourselves."

The best time to view outer planets orbiting exterior to the Earth is when they're near opposition and rise opposite to the setting Sun, and the best time to view the interior worlds of Mercury and Venus is when they're near greatest elongation and appear farthest from the Sun. The inner planets make a looping motion through either the dawn or dusk sky, never straying very far from the Sun.

MERCURY

The most elusive of the naked eye planets, Mercury never strays far from the Sun. Still, you can spy this fleeting world low in the dawn or dusk sky, if you know exactly where and when to look for it (see graphic, page 47). Not all apparitions of Mercury are favorable, though, due to two factors: one, Mercury has a noticeably elliptical orbit, meaning that elongations can vary anywhere from 17.9 degrees (near perihelion) to 27.8 degrees (near aphelion) from the Sun. Six to seven elongations can happen per year, alternating between dawn and dusk. Second is the angle of the ecliptic for any given time of the year versus observer latitude: for example, as seen from 40 degrees north near Philadelphia in March, the ecliptic is approaching perpendicular with the horizon at dusk. Reverse any of these factors (move the location south, or change the month or time from dusk to dawn), and the reverse is true, putting the ecliptic (and Mercury) in the weeds, low to the horizon.

"THE FAULT, DEAR BRUTUS, IS NOT IN OUR STARS, BUT IN OURSELVES."

At the eyepiece, Mercury displays a small, featureless disk, which waxes and wanes like the Moon. Mercury can range in brightness from magnitude -2.6 to +5.7. The disk of Mercury can range from 4.5 to 13 arcseconds in size, and appears half-illuminated near greatest elongation.

Mercury can transit the face of the Sun roughly once per decade. This occurs next on November 11, 2019, and again on November 13, 2032.

VENUS

The most brilliant natural object in the sky next to the Sun and Moon, Venus can actually cast a visible shadow and can be easily seen in the daytime near greatest elongation if you know exactly where to look for it. The greatest elongations for Venus can range from 45 to 47 degrees from the Sun, and the planet can reach a maximum brilliancy of magnitude -4.9. Apparitions of Venus nearly repeat their circumstances every 8 years, as thirteen orbits of Venus very nearly equal eight for the Earth.

At the eyepiece, Venus also shows discernible phases, and can range in apparent size from 66 arcseconds as a thin crescent passing between the Earth and Sun, to a near full phase, 9.7-arcsecond dot near solar conjunction on the opposite side of the Sun. This cycle of phases gave Galileo his first clue that Venus was truly fixed in its orbit around the Sun, and not the Earth.[8] A keen-eyed observer might just be able to pick out the horns of Venus as it passes through inferior conjunction, and it is possible to track Venus right through from the dusk to dawn sky on years when the tilt of its orbit tacks toward a maximum nearing 10 degrees from the Sun, as it will next approach in August 12, 2023, (8 degrees south) and again on March 22, 2025 (8 degrees north) . . . just be sure to attempt this observation when the Sun is still below the local horizon. Simply add the Venus rule of eight to 2023 and 2025 for subsequent wide solar conjunctions.

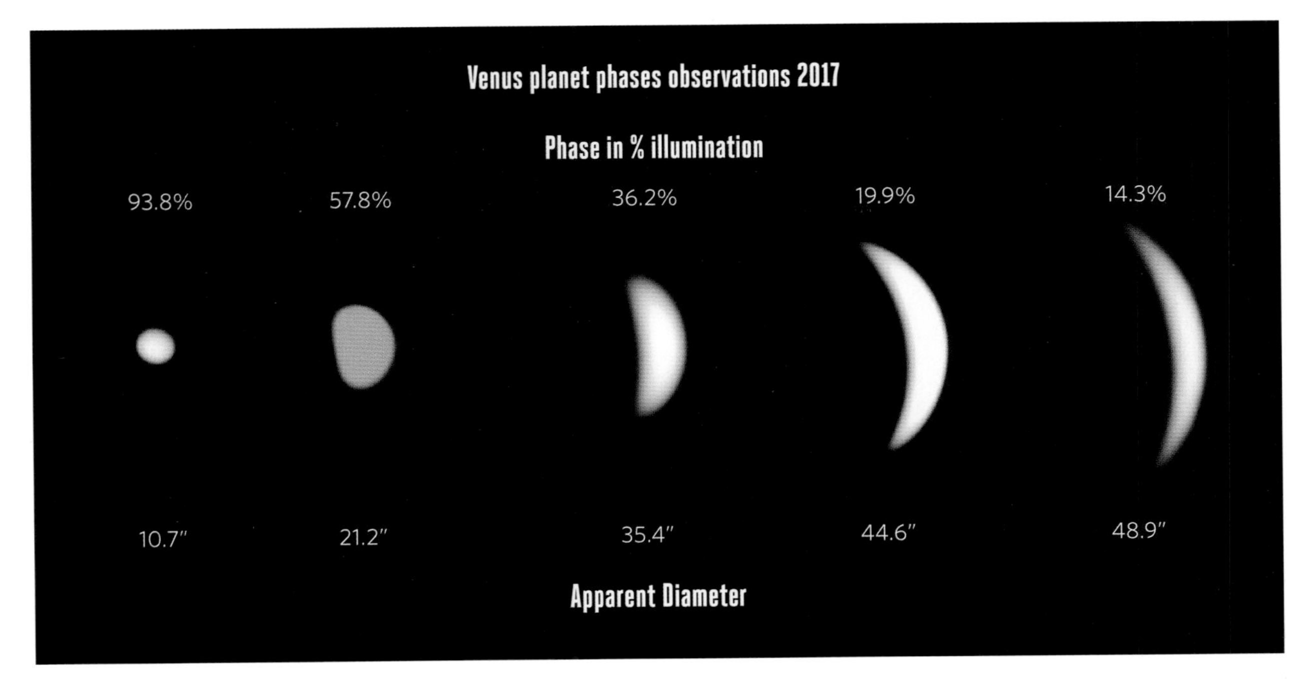

The phases of Venus as it approaches inferior conjunction between the Earth and the Sun.

The maximum elongation of Venus at inferior conjunction for the remainder of the twenty-first century is 8 degrees, 50 arcminutes on March 3, 2089.

Unfortunately, the disk of Venus is featureless in a telescopic view, as the world is perpetually shrouded in a noxious haze of carbon and sulfur dioxide. We always think it's a true cosmic irony that the nearest planet to us—which presents the largest disk—is also blank and featureless. Modern amateurs have, however, managed to tease out true detail from the Venusian cloud tops using ultraviolet filters.

Two observational mysteries of Venus remain. First, the observed date of **dichotomy**—when Venus appears to reach half phase—is often off by several days from the predicted date. Second, like the Moon, Venus often displays a reported ashen-light effect on its dark limb in crescent phase . . . but unlike our Moon, Venus has *no* convenient large reflector like the Earth nearby. Are these both optical illusions? Or maybe the result of air glow, lightning, or Venusian auroras? Whatever the cause, it's fun to remember that there are still mysteries in our very own Solar System, right next door.

MARS

Named after the Roman god of war, Mars can present a fascinating and changing view. Apparitions of Mars go through a 26-month cycle, and the disk of the planet can range from 3.5 arcseconds (at solar conjunction) to 25.1 arcseconds during a favorable opposition, the greatest range in apparent size for any planet. Oppositions of Mars are also variable, due to its elliptical orbit. This means that Mars can range anywhere from 25 to 14 arcseconds in size near opposition, and this follows a 15-year cycle.

Mars can range in brightness from magnitude +1.6 to -3, rivaling Jupiter and the brightest stars. Though it's often referred to as the Red Planet, Mars actually wears a range of shades from pumpkin orange to a sickly yellow, perhaps denoting that a planet-wide dust storm is underway.

The planet Mars nearing opposition.

Jupiter, along with its moon Io casting a shadow on the Jovian cloud tops. Stack of 400 images at 30 frames per second.

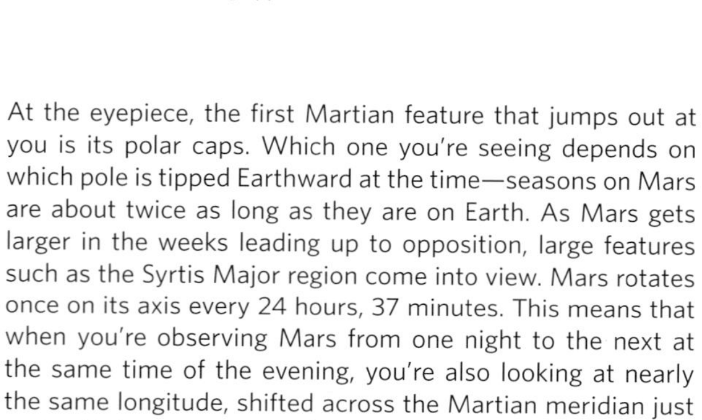

At the eyepiece, the first Martian feature that jumps out at you is its polar caps. Which one you're seeing depends on which pole is tipped Earthward at the time—seasons on Mars are about twice as long as they are on Earth. As Mars gets larger in the weeks leading up to opposition, large features such as the Syrtis Major region come into view. Mars rotates once on its axis every 24 hours, 37 minutes. This means that when you're observing Mars from one night to the next at the same time of the evening, you're also looking at nearly the same longitude, shifted across the Martian meridian just under 10 degrees westward.

The tiny Martian moons of Phobos and Deimos are tough, though not impossible targets during opposition. The best way to cross these off of your life list is to use an **occulting bar eyepiece**, or to place Mars just out of view.

Phobos shines at magnitude +11.3 and never strays farther than 20 arcseconds from Mars near opposition, and Deimos reaches magnitude +12.4 and reaches a maximum of 66 arcseconds from Mars at greatest elongation.

A great app to use to make sense of just what features you're seeing on Mars is Mars Previewer.

JUPITER

The largest planet in our Solar System, Jupiter reaches opposition once every 399 days, moving roughly one zodiacal constellation eastward per year. Jupiter shines a magnitude -3 and can show a disk up to 50 arcseconds wide at opposition.

At the eyepiece, the first feature that jumps out at you is the twin cloud belts striping the Jovian cloud tops, and the four large Galilean moons of Io, Europa, Ganymede, and Callisto. It's fun to watch the shadow play created by these moons, as they disappear into the shadow of Jupiter and cast shadows on its cloud tops. This shadow casting is nearly straight back near opposition, then off to one side near quadrature. Watch shadow transits of the Jovian moons long enough, and you can soon tell which moon is involved: innermost Io casts a small, inky-black shadow, while outermost Callisto casts a wide, grayish shadow. Callisto is also the only moon that can miss casting a shadow on Jupiter on some years. Double shadow transits of the moons are also neat to see, and occur in seasons several times a year. Rarer still are triple shadow transits, though you'll never see a *quadruple* transit, as the inner three moons are only locked in a 1-2-4 orbital resonance. The next triple shadow transit occurs on December 30, 2032.

At the eyepiece, you might manage to nab the famous **Great Red Spot** (GRS) of Jupiter, a massive swirling storm 1.3 times as wide as the Earth. The GRS has been rather lackluster in recent years, appearing more of a pale salmon color than truly red. There's also conjecture that the Great Red Spot may disappear altogether sometime this century. Jupiter rotates on its axis once every 10 hours, and near opposition you can follow it through one complete rotation in a single night.

Though the northern equatorial belt seems to be a permanent fixture, the southern equatorial belt is notorious for pulling a disappearing act roughly once a decade, for reasons unknown.

SATURN

Saturn is a true treat, the highlight of any star party. Shining at magnitude -0.2, Saturn takes 378 days to return to opposition. At the eyepiece, Saturn shows off its true glory, as its extensive system of rings is revealed even at low power. Even my old 60 mm refractor would readily show off the rings of Saturn, tiny and delicate. Crank up the magnification, and you might spy the **Cassini Gap**, a space between the A and B ring first documented by Giovanni Cassini in 1675. The rings also appear to tip from edge on to wide open to edge on again, following a 29.5-year cycle. Also watch for the globe of Saturn casting its massive shadow back on the rings. This adds an especially 3-D effect to the view of the planet, right around quadrature.

Shining at magnitude +8.5, you might just be able to spy Saturn's largest moon Titan with binoculars. Titan is the second-largest moon in the Solar System, next to Jupiter's moon Ganymede. Five other moons of Saturn are within range of a backyard telescope: +11 magnitude Enceladus, +10.2 magnitude Tethys, +10 magnitude Rhea, +10.4 magnitude Dione, and Iapetus. Keep an eye on Iapetus, which varies in brightness from magnitude +10.2 to +11.9, as it presents two drastically different hemispheres toward us once every 79-day orbit.

MORE TINY WORLDS BECKON EVEN FARTHER OUT.

URANUS

Shining at magnitude +5.5, the blue-green ice giant Uranus reaches opposition once every 370 days. Its 4-arcsecond disk gives it away in the starry field as a pseudo-star that refuses to snap into focus. In the case of the outer planets, just knowing what you're seeing is half the fun.

NEPTUNE

More of a gray-blue +8th magnitude dot, Neptune presents a 2.3-arcsecond disk and reaches opposition once every 367 days. Keen-eyed observers with large light bucket telescopes may want to try for Neptune's large +13 magnitude moon Triton, which reaches a 15-arcsecond elongation once every 3 days.

Orbiting the Sun once every 165 years, Neptune just completed its first orbit of the Sun, since its discovery in 1846, in 2011.

PLUTO AND MORE

Discovered in 1930, Pluto shines at magnitude +14 and orbits the Sun once every 248 years . . . and for decades, that's about all anyone really knew about this enigmatic world. NASA's New Horizons spacecraft gave us our first real views of Pluto and its large moon Charon during its July 2015 flyby (see image, page 49). The act of finding Pluto requires a good finder chart and rewards you with knowing that the faint star is indeed the controversial dwarf planet. Over the next decade, Pluto will slowly depart the apparent direction in the sky along the galactic plane as it traverses the constellation Sagittarius until it enters Capricornus in 2024.

More tiny worlds beckon even farther out. Though they're relative newcomers, amateurs including Mike Weasner have managed to image dwarf planets Makemake (+17 magnitude) and +17th magnitude Haumea. Eris (+18th magnitude) should also be within range of the determined amateur.

ACTIVITY: REPORTING OBSERVED CHANGES ON OTHER PLANETS

You're a skilled observer, checking out a target planet at high magnification. Perhaps you're sketching what you see, (see our handy activity Sketching the Moon and Planets on page 55), maybe you're imaging said distant world, or maybe you're just out causally observing.

Does something look out of place to you?

When it comes to amateur astronomy, it's always worth keeping an eye out for the surreptitious. Believe it or not, amateur astronomers still make discoveries in our very own Solar System. Time on professional telescopes is precious and sought-after. Typically, telescopes like the Hubble are busy looking at remote galaxies, quasars, or other cutting-edge astrophysical objects. They're often only repurposed to look at planets after a change or event is announced, such as the crash of Comet Shoemaker-Levy 9 into Jupiter in July 1994.

What professional astronomers *aren't* doing tonight is looking through the eyepiece at a blob of a planet, pipe in one hand, focuser in the other. Today's professional astronomer is often instead looking at a computer screen, perhaps remotely from the warmth of their university office, thousands of miles from the observatory. Often, it's keen-eyed amateurs that raise the alarm when new action occurs on a planet.

Watching for changes on Solar System bodies is a fun activity, and hey, you just might be the first one to spy a breaking event . . . but how do you tell the world? The clearing house for amateur observations is the Association of Lunar & Planetary Observers (ALPO) (http://alpo-astronomy.org/). I also suggest that serious observers join the numerous Yahoo! Discussion groups listed on the ALPO's home page. There's always a good ongoing cross-talk discussion on these pages about just what observers are currently seeing, and folks are always willing to confirm observations worldwide.

Below is a brief rundown of typical changes to watch out for.

The Moon

Usually considered a dead world, **Transient Lunar Phenomena** (TLPs) have long been reported, even by skilled observers. These usually take the form of spurious brightness changes or even discolorations of regions. Two parts of the Moon notorious for TLP reports are the Aristarchus and Kepler craters.

Mars

The onset of Martian dust storm season is often first announced by amateur astronomers. In March 2012, the amateur astronomer Wayne Jaeschke imaged a strange cloud on the limb of Mars.

Jupiter and Saturn

Storms roil and new ones periodically form and disperse along the cloud tops of these gas giants.

Other Changes to Look Out For

The ALPO also tracks cometary outbursts, and it's always worth watching for impacts on any Solar System body, including the Moon . . . you just never know what you might see if you watch long enough.

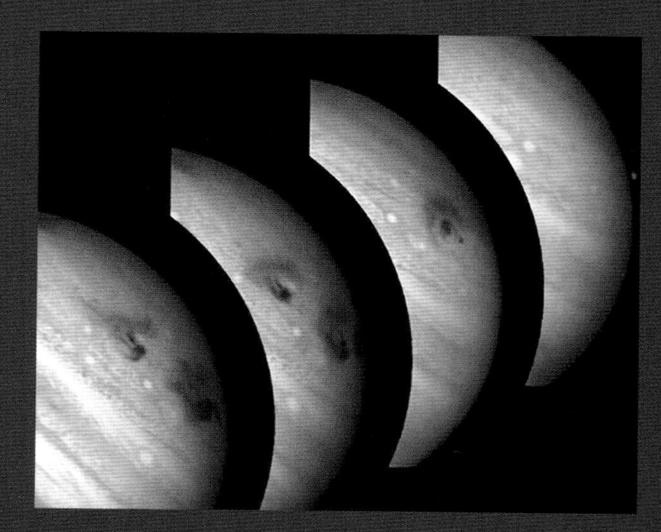

The July 1994 impact of Comet Shoemaker-Levy 9 on Jupiter, captured by the Hubble Space Telescope.

OBSERVING ECLIPSES

Eclipses are one of the most wondrous events the sky has to offer. These alignments of the Sun, Earth, and Moon are magnificent celestial spectacles not to be missed. Eclipses come in two types: lunar and solar. And they're not as rare as you might think; you only have to be on the correct hemisphere of the Earth to see a lunar eclipse, and partial solar eclipses visit any given continent nearly every year. Seeing the rarest spectacle of all—a total solar eclipse—often requires careful planning years in advance to make the journey to the narrow path of totality.

A partially eclipsed Sun rises over the Vehicle Assembly Building at the Kennedy Space Center on the Florida Space Coast on the morning of November 3, 2013.

Seven to four eclipses can occur in a given calendar year. The Moon's orbit is inclined 5 degrees with respect to the plane of the ecliptic, the imaginary plane of the Earth's orbit traced out across the celestial sphere. If the Moon's orbit wasn't inclined to the ecliptic, we would see two eclipses every lunar month, one solar and one lunar. Instead, we only see eclipses when the nodes where the intersection of the Moon's orbit and the ecliptic align with the Sun and Earth. This alignment is known as a **syzygy**, an ultimate Scrabble word (though you need a blank tile for the third "y") to land on a triple word score.

Moreover, the ascending and descending nodes move around the ecliptic once every 8.9 years in what's known as the **precession of the line of apsides**. This motion is due mainly to the Sun's gravity, dragging the Moon's orbit around the Earth. This means that eclipses are paired together, with lunar and solar eclipses occurring about two weeks apart in **eclipse seasons**.

Another key cycle to unlocking the math underlying eclipses is the **saros cycle**. Two hundred and twenty-three lunations (18 years, 11 days, and 8 hours) brings the Sun, Moon, and the Earth back to very nearly the same geometry. This means that one saros after a given eclipse—lunar or solar—an eclipse with very similar circumstances will occur, shifted 120 degrees longitude westward. Eclipses belong to saros families, spanning millennia.

It's amazing to think that ancient cultures such as the Babylonians unlocked the secrets of the saros and could predict eclipses, and they did this using nothing more than local historical records, naked eye observations, and the power of math.

Solar eclipses, when the Moon passes in front of the Sun as seen from the Earth, come in four varieties: **partial**, **annular**, **hybrid**, and **total**. Partials occur when the Moon only partially obscures the Sun.

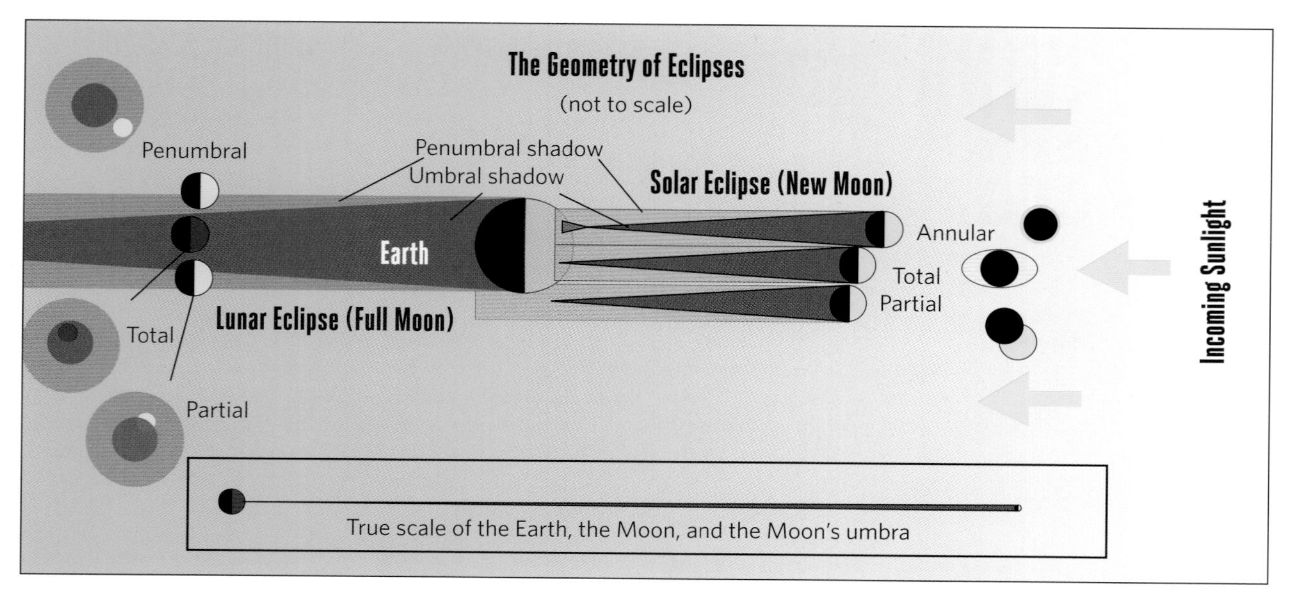

The geometry of lunar versus solar eclipses and the different resulting types of eclipses. When the Moon passes between the Earth and the Sun, a solar eclipse occurs (right). If the dark inner shadow or umbra of the Moon reaches the Earth, a total solar eclipse occurs along its path. If the umbra misses and the Earth just grazes the outer lighter edge (the penumbra) of the Moon's shadow, a partial solar eclipse occurs. If the Moon's umbral shadow fails to reach the Earth, an annular eclipse occurs. Likewise, a total lunar eclipse always occurs near a Full Moon, if the inclined path of the Moon crosses the ecliptic at an ascending or descending node. A lunar eclipse can also be total, partial, or penumbral, depending on how central the Moon's passage is through the Earth's shadow.

An annular eclipse (technically a special type of partial eclipse) takes place when the Moon is too visually small to cover the Sun. The Moon and Earth both have orbits that are slightly elliptical, and an annular eclipse happens when the Moon is near apogee, or its most distant point from the Earth. The resulting bright ring of fire seen during an annular eclipse is known as an **annulus**, and the core shadow path for an annular eclipse is known as the **antumbra**.

A total solar eclipse is the most spectacular of them all. It's a bizarre spectacle, to witness a false dawn fall upon the landscape at midday. The temperature drops, roosters may begin to crow, and ethereal shadow bands flit across the dimming landscape. The path of totality for a total solar eclipse is generally dozens of kilometers wide, and totality can range anywhere from near-instantaneous to 7 minutes, 32 seconds in duration as the shadow of the Moon races across the Earth's surface from west to east at up to 1.4 kilometers per second.

A hybrid solar eclipse is simply a combination of an annular and total solar eclipse, with the central path of the eclipse annular along one stretch, and total along another.

Totality! The view from Casper, Wyoming, during the August 21, 2017 total solar eclipse.

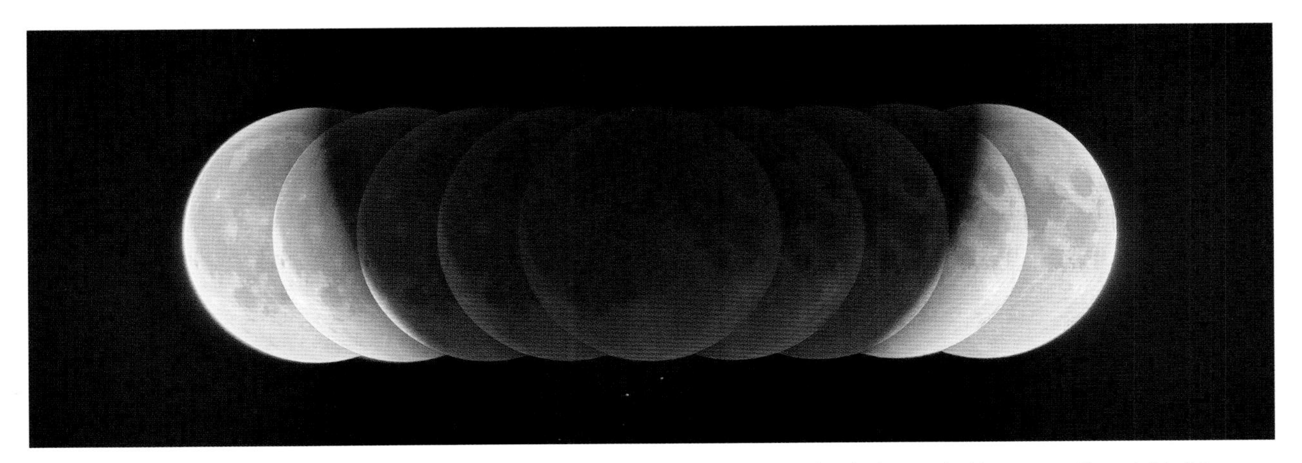

The stages of a total lunar eclipse. Note that you can actually see the curve and scale of the Earth's shadow as the Moon passes through it in this fine compilation, taken on January 31, 2018. Taken using a Sony A7m2, Contax camera, ISO 320-6400, multiple exposure, merged in Photoshop, sky blended using layer mask in Photoshop.

Phenomena to watch for during a total solar eclipse include **Baily's Beads** and the **Diamond Ring effect**, as sunlight streams through lunar valleys and the pearly white corona of the Sun emerges, a sight only visible during totality. Large prominences leaping off through the chromosphere of the Sun may also be seen.

It's a wonderful coincidence of our current epoch: The Sun is 400 times larger in diameter than the Moon, but it's also roughly 400 times farther away, making the two appear roughly the same size in the sky. Of course, this is just a generalized statement of the true situation; over the five-millennia span from 2000 BC to 3000 AD, annular solar eclipses are slightly *more* common than totals (that is, 33.2 percent of all solar eclipses for the current five millennia span are annular, versus 26.7 percent totals—partials and hybrids make up the remainder).

And this situation is slowly changing, as the Moon recedes away from us at a current rate of 3.8 centimeters a year. A little under a billion years ago, the first brief annular eclipse occurred, unseen by human eyes. Likewise, just under 600 million years from now,[9] totality will grace the skies of Earth, one last time.

Safety precautions must be used during all partial phases of a solar eclipse. The same is true for an annular eclipse. Remember, 1 percent of the Sun is still *pretty* bright, enough to cause eye damage. Only use filters such as Baader AstroSolar Safety Film or approved glass filters that secure firmly to the front of a telescope specifically designed for solar viewing. NEVER use filters that screw directly onto the eyepiece, as these can overheat and crack.

Projection is another safe method to observe a solar eclipse. You can use a simple shoebox pinhole projector, or a telescope or pair of binoculars to project the Sun onto a white piece of paper. Make sure you attend to your telescope at all times while it's aimed at the Sun (kids and even adults *will* still try to peek in the eyepiece), and be sure to remove and stow your finderscope, or cover it up. Also, telescope optics—especially those found in Schmidt-Cassegrains—can heat up pretty quickly when projecting the Sun, causing the adhesives in the lens elements to come undone, or possibly warping or cracking a lens or a mirror. There have even been documented cases of unattended telescopes focusing the Sun and starting fires. Be sure to only use the projection method for a minute at a time, then allow the telescope to cool off for a few minutes.

Lunar eclipses are much more casual affairs. These occur when the Moon is opposite to the Sun, with the Earth between the two. A lunar eclipse can be **penumbral**, **partial**, or **total** and always occurs when the Moon reaches Full phase.

A penumbral lunar eclipse occurs when the Moon misses the inner dark umbral shadow of the Earth, and instead passes through the lighter outer penumbral cone. Standing on the Earthward face of the Moon looking back at the Earth, you'd see a partial solar eclipse.

A partial lunar eclipse occurs when the Moon grazes the dark inner umbral shadow of the Earth. The Earth's shadow is almost three times as large as the Moon at the Moon's distance of a quarter million miles away. This is one of the few times you can actually *see* direct evidence that the Earth is indeed round, as the curve of the planet's shadow is cast across the disk of the Moon.

A total lunar eclipse is spectacular, and totality can last a maximum of nearly 109 minutes. Watch for the Moon to turn blood red once it's immersed in the umbra. You're seeing the filtered light from a thousand sunsets worldwide, as sunlight streams through the atmosphere of the Earth, back onto the Moon.

All total lunar eclipses are not equal. Watch the color of the Moon during totality, as it can take on anything from a dark brick color to sickly deep yellowish cast. This change is due to the amount of dust and aerosols present in the Earth's atmosphere. When volcanic eruptions occur, the Moon typically appears a dark, deep red. For example, the Moon nearly disappeared during a total lunar eclipse after the eruption of Mount Pinatubo in 1992!

The color of the eclipsed Moon is described by its **Danjon number**, with 4 being very bright and 0 being exceptionally dark.

THE SCIENCE OF ECLIPSES

Solar eclipses play a key role in the history of astronomy. Astronomers first detected the element helium in the spectra of the Sun's corona during an eclipse in the late nineteenth century. An expedition led to a small island in the Southern Atlantic by Sir Arthur Eddington in 1919 made the first great observational proof of Einstein's theory of general relativity, as the light of a distant star was deflected passing near the mass of the Sun during totality.

Today, eclipses are more a source of beauty than science. Eclipses are still useful, however, in efforts to make highly accurate measurements of the size of the Sun and measuring the diameter and the fuzzy edge of the Earth's shadow as seen at the Moon's distance.

How rare are total solar eclipses in the Universe? Is the celestial situation enjoyed by humanity a solitary occurrence in the Milky Way Galaxy? Well, it's tough to say just what alien eyes might witness from the surface of their home worlds, but the large moons of Jupiter do indeed witness an almost precise similar fit, as one moon crosses in front of the other. These are, however, much faster events, with totality measured in tens of seconds at most.

Perhaps some future human astronaut or colonist will stand on the surface of Callisto or Ganymede and witness such a scene. For now, though, we can gaze in wonder at the cosmic dance of eclipses as they cross Earthly skies.

HANDY ECLIPSE RESOURCES

Eclipse Maps (maintained by the eclipse chaser and umbraphile Michael Zeiler):
http://eclipse-maps.com/Eclipse-Maps/Welcome.html

Eclipse Weather (maps and weather predictions for future solar eclipses):
http://eclipsophile.com/

NASA/Goddard Space Flight Center's Eclipse Web page (maintained by Fred Espenak):
https://eclipse.gsfc.nasa.gov/eclipse.html

FUTURE ECLIPSES OVER THE NEXT 6-YEAR SPAN (2019–2024)		
Date	**Type**	**Region Favored**
January 6, 2019	Partial solar	Aleutian Islands, Northeast Asia, northern Pacific
January 21, 2019	Total lunar	Central Pacific, Americas, Europe
July 2, 2019	Total solar	Southern Pacific, South America
July 16, 2019	Partial lunar	South America, Europe, Africa, Asia, Australia
December 26, 2019	Annular solar	Southeast Asia, Australia
January 10, 2020	Penumbral lunar	Europe, Asia, Africa, Australia
June 5, 2020	Penumbral lunar	Europe, Asia, Africa, Australia
June 21, 2020	Annular solar	Northeast Africa, Southern Asia
July 5, 2020	Penumbral lunar	Americas, Southwest Europe, Africa
November 30, 2020	Penumbral lunar	Asia, Australia, Pacific, Americas
December 14, 2020	Total solar	Southern South America
May 26, 2021	Total lunar	East Asia, Australia, Pacific, Americas
June 10, 2021	Annular solar	Europe, Northern Asia, Arctic
November 19, 2021	Partial lunar	Americas, Northern Europe, East Asia, Australia, Pacific
December 4, 2021	Total solar	Antarctica, South Africa
April 30, 2022	Partial solar	Chile, Argentina
May 16, 2022	Total lunar	Americas, Europe, Africa
October 25, 2022	Partial solar	Europe, Middle East
November 8, 2022	Total lunar	Asia, Australia, Pacific, Americas
April 20, 2023	Hybrid solar	Southeast Asia/Indonesia
May 5, 2023	Penumbral lunar	Australia, East Asia
October 14, 2023	Annular solar	Southwest USA, Central and South America
October 28, 2023	Partial lunar	Africa, Asia, Europe
March 25, 2024	Penumbral lunar	Americas
April 8, 2024	Total solar	Mexico, USA, Eastern Canada
September 18, 2024	Penumbral lunar	Atlantic region
October 2, 2024	Annular solar	Southern tip of South America

OCCULTATIONS:

THE FINE ART OF WATCHING ONE THING PASS IN FRONT OF ANOTHER

The sky often gives us hints to its true nature, in simple ways that have terrestrial analogs. For example, watch an aircraft as it passes through the sky on a sunny day. Does it pass in front of or behind an occasional stray cloud? From this simple observation, we can deduce which is closer to us.

This is simple, I know, but celestial bodies do the same thing, and this casual motion of one body passing in front of another gave early skywatchers the first inkling that the vault of the sky is not truly fixed, but exists instead with some objects closer, and some farther away.

This sort of event leads us to one of the more head-turning terms in modern astronomy, straight out of its hoary roots with astrology: an **occultation**. Try it: Use this term with some of your non-astronomy friends sometime soon ("I watched the Moon occult Venus last night . . .") and this will confirm their secret hunch you're actually dabbling in astrology, despite your protests at the term (the very best way to anger an astronomer is to call them an astrologer).

We get the term "occlude" from the same root as "occult," meaning to block or cover something up. When a smaller body crosses the disk of another—say, the planet Mercury moves across the disk of the Sun—we call this a **transit**. Technically, a total solar eclipse is an occultation of the Sun by the Moon, and an annular solar eclipse is a transit of the Moon across the solar disk, though it's fairly pedantic to refer to them as such. Eclipses are such singular and notable spectacles that they warrant their very own definition, though yes, the terms "eclipse" and "occultation" can be used fairly interchangeably. Like the term "planet," many terms in astronomy are cultural definitions, again, betraying the science's ancient connections with the humanities.

The waning crescent Moon, just prior to occulting the planet Jupiter and the four Galilean moons.

The most convenient occulting body we have from our Earthly vantage point is our single large Moon. Passing across the starry vault of the sky once every sidereal month (27.3 days), it cuts a 30-arcminute-wide (half an angular degree) path, blocking every object in its way. When the Moon is waxing through the evening sky, its dark edge leads the way, with its bright sunlit side trailing. When the Moon passes Full and begins waning toward New, however, the opposite is true, with the bright daytime edge of the Moon's limb leading the way, and stars or planets reappearing from behind the dark trailing limb.

And like everything else, there's a term for that disappearance and reappearance of objects from behind another during an occultation: We call the disappearance **ingress**, and the reappearance **egress**.

Occultations are abrupt, split-second events in a Universe that otherwise moves at a sometimes glacial pace. Double stars may dance in an orbit lasting centuries. The sky will look pretty much the same to you tonight as it looked to your grandparents, or will to your grandchildren. The Sun takes a quarter of a billion years to orbit around the core of the Milky Way Galaxy, longer than there have been Homo sapiens calling the Earth home.

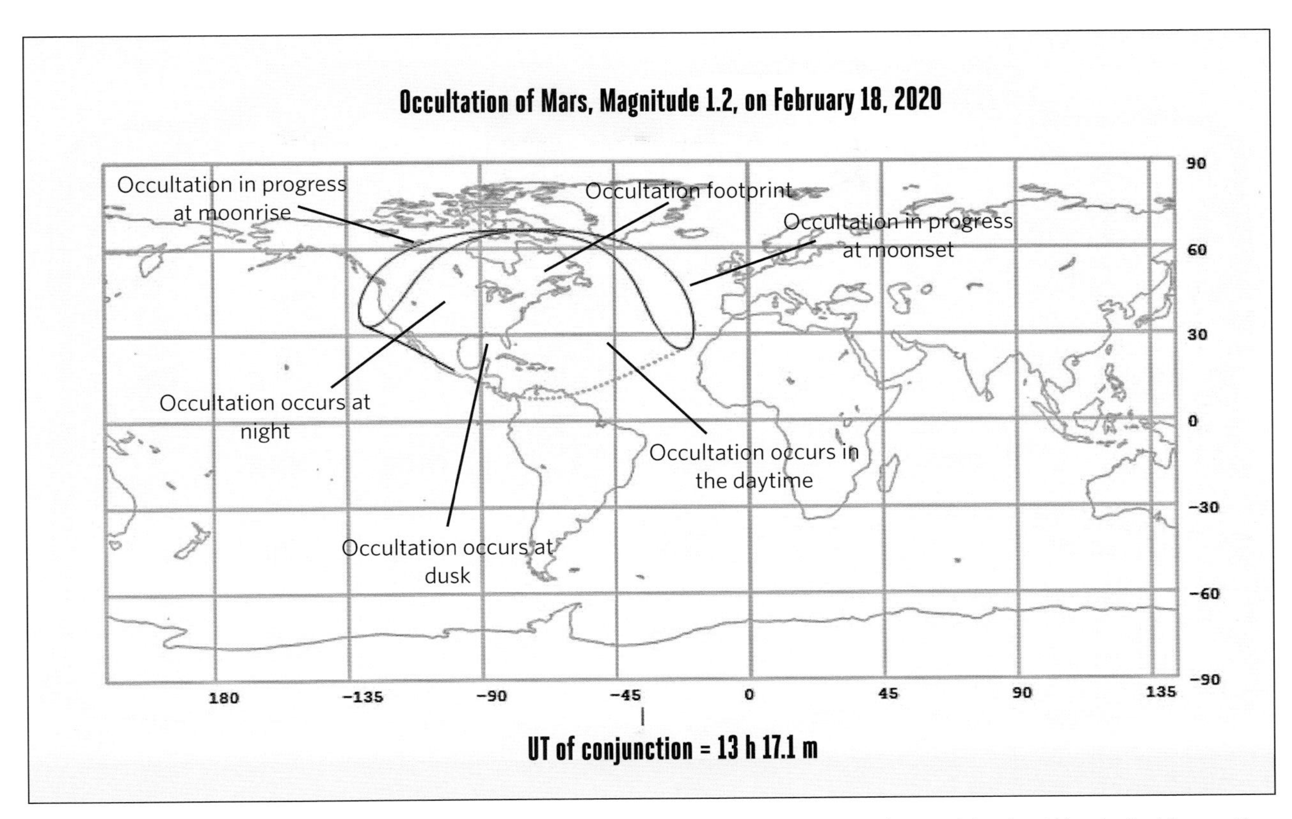

A typical lunar occultation footprint across the surface of the Earth. In this case, the map depicts an occultation of the planet Mars by the Moon on the morning of February 18, 2020, over North America. The dashed lines mark the edge of the area from which the occultation occurs during the daytime, and the solid boundary marks the area where the occultation occurs under dark skies. The "bent figure eight" area marks where the occultation is in progress while the Moon is either rising or setting.

But an occultation is over—often in a split second—as the Moon interrupts light from a distant star in the very last second before those photons reach the back of your eyeball and nervous impulses can transmit them to your brain for interpretation as light, or the lack thereof. Those photons were forged by nuclear processes in the heart of a distant star, then bounced around for thousands of years in its super-dense core like a ping pong ball until finally entering open space, shining free at the speed of light for centuries, before getting photobombed by the slag pile that is Earth's Moon in the very last second.

With an orbit inclined 5 degrees relative to the ecliptic plane, the Moon can occult all five classical naked eye planets, and four +1st magnitude stars in our current epoch: Antares in the astronomical constellation Scorpius, Aldebaran in Taurus, Regulus in Leo, and Spica in Virgo. Antares is also the most frequently occulted, with 386 events for the current century running out to 2100 AD.

Since the Moon's path is fixed with respect to the ecliptic plane and not the Earth's rotational equator, these bright star occultations occur in cycles. When the Moon's path is shallow relative to the celestial equator, many occultations of stars and planets are seen; when the Moon's path becomes steep and hilly, occultations of objects along the ecliptic become fewer and less frequent, as the Moon seems to plunge down and up again across the plane of the Solar System once every sidereal month.

But once a star is in the Moon's crosshairs, a cycle of occultations begins, and occurs once for every orbit. The footprint of where these events are visible on the face of the Earth evolves with each pass, and like eclipse saros cycles, they start toward one pole, progressing slowly toward the equator as they become more central, varying in lunar phase as well. Generally through an occultation cycle, at least one or two events will favor a given location on Earth. The most dramatic occultations occur when an object ingresses on the dark limb of a waxing crescent Moon, while occultations near Full phase can be tough, owing to the glare of the Moon.

You can watch an occultation of a bright star by the Moon with the naked eye, no equipment necessary. We've even managed to catch occultations of bright stars and planets near the daytime Moon using binoculars or a telescope.

We say "four bright stars can currently be occulted by the Moon in the current epoch" because this situation is slowly changing, mainly due to our friend, the 26,000-year-long wobble of Earth's axis known as precession of the equinoxes, and the individual proper motions of the stars themselves, carrying them slowly into and out of the Moon's way. In fact, until 117 BC[10]—the blink of a celestial eye—the Moon could also occult the bright star Pollux in the constellation Gemini.

Planetary occultations are fun to watch as well. We can actually see evidence for planets as tiny disks during occultations with the naked eye, as it takes a leisurely few seconds for the Moon to cover them up. During an occultation of Venus, say, you can see a crescent Moon run over a crescent planet, an event you might just witness in the daytime sky. This sort of dramatic occultation is represented by the star and crescent adorning the flags of many Islamic nations, and is said to have been inspired by an occultation of a star or planet by the Moon seen by Sultan Alp Arslan in August 26, 1071, shortly around the time of the defeat of the Byzantine Army at the Battle of Manzikert.

Aldebaran Occultation—the gibbous Moon versus the bright star Aldebaran (off of the Moon's lower right) near occultation.

And as with eclipses, we can use references to occultations to gain a fixed point for a given historical text that might mention one, as well. One great example is the occultation of Jupiter by the Moon on January 17, 121 BC, perhaps commemorated on a Greek coin from the same era.[11]

Occultations would've even made great timepieces for ships at sea . . . if only the pesky Moon would stay put. At a quarter of a million miles distant, the Moon is close enough to display parallax from different vantage points on the surface of the Earth. This is why an occultation might be visible from one continent or hemisphere, but not another.

Occultations of Mercury or Venus always occur near the Sun at waxing or waning phase for the Moon, while occultations of the outer planets can occur at any phase. Occultations of Mercury by the Moon are toughest of all, as it never strays farther away from the Sun than 28 degrees. In fact, occultations less than 10 degrees from the Sun are for all intents and purposes impossible (and unsafe) to observe.

Rarest of all are double occultations, where the Moon covers up a naked eye pair of planets, or a planet and a bright star simultaneously. These astronomical curiosities are truly once-in-a-lifetime events, occurring over narrow (and usually remote) stretches of land or water. This last such occurrence was when the Moon occulted Venus and Jupiter as seen from the Atlantic island of Ascension on April 23, 1998, and the next one isn't until the Moon covers up Regulus and Venus (possibly a grazing event for Regulus) as seen from a remote patch of Siberia on September 19, 2025.

The Moon moves roughly its own apparent 30-arcminute (half a degree) diameter once per hour, meaning any close double occultation must occur about an hour apart . . . one might witness an emoticon "smiley face" triple grouping during such an event, evidence that perhaps the Universe does indeed have a sense of humor.

Also rare are instances where a planet might pass in front of another, or a bright star. Planets are small enough in apparent size that these also occur only a few times per century. For example, Venus occulted the bright star Regulus on July 7, 1959, and will do so again on October 1, 2044. And if you're reading this in 2065, watch for a fine occultation/transit of Venus in front of Jupiter on November 22, 2065.

And speaking of the far future, any astronauts looking back at the Earth as the Moon occults Mars would see a fine transit of the gray Moon across the verdant disk of the Earth.

At the telescope, we can see the four bright Galilean moons of Jupiter (from innermost to outermost: Io, Europa, Ganymede, and Callisto) as they transit the face of Jupiter once per orbit and disappear in the giant gas giant's shadow. These events actually allowed the eighteenth-century Danish astronomer Ole Rømer to measure the speed of light, as he realized that discrepancies between predicted and actual disappearances of the moons were due to variations in the distance between the Earth and Jupiter. Seventeenth- and eighteenth-century navigators also hoped to use the phenomena of Jupiter's moons to get a one-time fix on the time and with it, a good gauge of longitude at sea. This is tough enough to do in practice, however, from a stable telescope on shore, let alone the deck of a pitching ship!

Another type of occultation has come into vogue in the past few decades: Asteroid occultations, when an asteroid blocks out a distant star and the shadow on the tiny space rock crosses over the Earth. One reason these are gaining in popularity is that the computing power to precisely measure asteroid orbits and knowledge of precise stellar position has only recently come of age with astrometry missions such as Hipparcos and the European Space Agency's Gaia mission, launched in 2013. Prior to 1980, less than a handful of asteroid occultations of stars had ever successfully been observed, period. Today, dozens of asteroid occultations are predicted and observed every month worldwide. Most require a telescope, although once or twice per year, an asteroid might cover up a naked eye star, such as the much-anticipated occultation of Regulus by 163 Erigone on March 20, 2014, which was unfortunately clouded out.[12]

WELCOME TO THE ADDICTING AND OCCULT ART

of watching one celestial object passing in front of another.

We've always wondered if, far in the unrecorded past, a nomadic hunter might've stopped, gazed at a bright star, and saw it wink out for a few seconds, then back on again, leaving them to wonder what *that* was all about . . .

One optical illusion that can occasionally be seen during a lunar occultation is known as the **Coleridge Effect**, as a star or planet seems to briefly hang between the two tips of the crescent Moon's horns, in seeming defiance of celestial mechanics. This comes from a reference in Samuel Coleridge's *Rime of the Ancient Mariner*, when the doomed protagonist notes seeing ". . . one bright star, almost atween the tips" of the "horned Moon."

Welcome to the addicting and occult art of watching one celestial object passing in front of another.

Check out Chapter 11: Real Science You Can Do and Protecting the Night Sky (page 198) for observing, recording, and contributing to astronomical occultation science.

THE DEEP SKY

Looking out at nebulae, star clusters, galaxies, and more.

"Vastness! And Age . . . Silence! And Desolation! And dim Night! I feel ye now—I feel ye in your strength . . . ever drew down from out the quiet stars."

— *Edgar Allan Poe*, The Coliseum

WISHING ON A STAR:

WHAT'S IN A NAME?

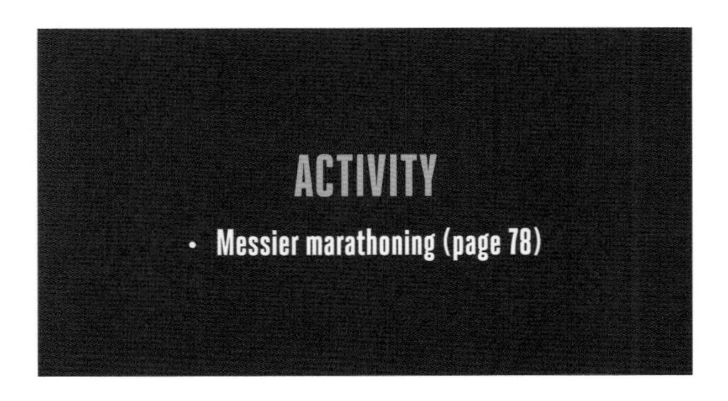

THE BABYLONIANS AND ANCIENT EGYPTIANS

saw basically the same sky you can see overhead tonight . . .

Stars are the signposts in the road map of the deep sky, with constellations as the territories. The ancient vault of the sky provides a fine nighttime backdrop for the drama of Earthly affairs. The Babylonians and the ancient Egyptians saw basically the same sky you can see overhead tonight,[1] and our descendants will continue to see the same starry vault overhead for thousands of years to come.

Early in 2017, the International Astronomical Union formalized the proper names of 212 stars.[2] These names come down to us from Arabic astronomers and have such cryptic monikers as Spica ("ear of wheat") and Betelgeuse (which every American now pronounces, largely thanks to the movie, as "Beetlejuice"). The poetic history of astronomy assures that we have Roman names for planets, crossing through mostly Greek constellations, adorned with Arabic stars. And though more obscure names such as Alhena and Mirzam have fallen into disuse, they seem to have made a minor comeback thanks to planetarium programs and GoTo telescope auto guide systems imploring users to "ALIGN ACHERNAR" in red glowing letters.

A more systematic and descriptive way to designate a star is by its order in brightness in a given constellation using the Latin alphabet, with alpha as the brightest, then beta, then gamma, and so on, along with the Latin genitive form of the constellation's name.[3] For example, you can also call Regulus "Alpha Leonis," denoting its title as the brightest star in the constellation Leo.

Stars also have aliases in the dozens of additional star survey catalogs, including the HD (Henry Draper), SAO (Smithsonian Astrophysical Observatory), and HIP (Hipparcos) catalogs. It can get quite confusing, I know, to have to cross reference the SAO designation of a star in one program, versus its HD name in another . . . as is often the case these days, the magic eight ball of Google comes to the rescue to help discern and cross index the many names of individual stars.

Left corner: The Rosette Nebula complex, a giant molecular cloud in the constellation Monoceros, the Unicorn, about 5,200 light-years away. Note from the photographer Dustin Gibson that this image was: "processed to leave all three channels in to be original Hubble SHO (sulfur=red, hydrogen=green, oxygen=blue) mix."

DEEP SKY OBJECTS:

AN INTRODUCTION TO THE MESSIER AND NGC CATALOGS

Over the last few centuries, astronomers have compiled the diverse menagerie of deep sky objects into the various collections and catalogs we use today. Of course, our knowledge of exactly *what* these fixed fuzzy patches in the sky are has changed over time. Some of these objects were bright enough to be visible in the pre-telescopic era, but most awaited discovery after the telescope was turned skyward.

MESSIER'S CATALOG

The first serious, extensive attempt to compile an observer's list of deep sky objects,[4] Charles Messier's catalog is a compendium of 110 objects covering the northern hemisphere sky down to declination -34.8 degrees south (that's Messier 7). The (possibly apocryphal) story usually told is that the eighteenth-century astronomer Charles Messier (born 1730, died 1817)[5] got tired of stumbling over these fixed fuzzy patches in the sky on his quest to discover new comets, and decided to do something about it. To this end, Messier compiled an observer's guide to these fixed false comets of the sky. Messier started this work in the early 1770s, and in true eighteenth-century fashion, Messier's list was completed in batches over almost a decade from 1774 to 1781 while he patiently scanned the sky with his 4-inch (10-cm) refractor.

Messier also left some mysteries for us to ponder. Why, for example, did he include large open clusters such as the Beehive (M44) and the Pleiades (M45)? These are very un-comet looking, and would seem to fly against the "here be fake comets" hypothesis that's often ascribed as his original motivation. Perhaps he merely included large clusters for the sake of completion . . . but why, then, did he miss the magnificent Double Cluster in Perseus? Why does his catalog include one double star (Messier 40),[6] and what exactly was Messier M73, often ascribed to a loose asterism of stars in Aquarius?

A portrait of the 18th-century French astronomer Charles Messier.

ENTER THE NGC CATALOG

The second-tier deep space object compilation is the *New General Catalogue of Deep-Sky Objects*, usually abbreviated as the New General Catalogue, or NGC for short. John Dreyer compiled the NGC list as we know it today in 1888, and the catalog traces its origins back to Sir William Herschel's son John and his *General Catalog of Northern and Southern Deep Sky Objects*, collected during several expeditions to South Africa to observe the southern hemisphere sky. More thorough and definitive than the Messier catalog thanks to the better optics that were available a generation later, the NCG catalog covers 7,840 objects down to about magnitude +16.5.

The NGC catalog offers observers a lifetime of deep sky targets to chase after.

The Barnard dark nebula patch through the constellation of Taurus, the Bull.

OTHER DEEP SKY CATALOGS

Not enough? There's a universe of deep sky objects out there for the determined amateur astronomer with access to a huge light bucket telescope and a dark sky. There is also a bewildering array of lesser catalogs of deep sky objects out there: ever heard of (or tried to pronounce) the Dolidze-Dzimselejsvili (DoDz) Open Clusters Catalog? As with stars and star designations, these catalogs are threatening to grow in size and membership, as new Earth and space-based surveys such as the Large Synoptic Survey Telescope (LSST), which is set to go online in 2019,[7] and the James Webb Space Telescope (JWST), which is launching in 2020, will push the magnitude envelope back ever fainter.

Some other deep sky catalogs in the grasp of mere mortals are listed below.

The Abell Catalog: An all-sky catalog of 4,073 galaxy clusters that was by completed George Abell in 1958 and revised in 1989.

The Arp Atlas of Peculiar Galaxies: Halton Arp published this catalog of 338 galaxies in 1966.

Barnard's Catalog of Dark Nebulae: A list of intriguing dark cloud objects along the plane of the Milky Way Galaxy. Prolific astronomer E.E. Barnard compiled this list, which contains such evocative objects as the Cone, Snake, and Coalsack Nebula.

The Caldwell Catalog: Sir Patrick Moore compiled this list of 109 objects, which complements and fills in many of the gaps in Messier's catalog as a sort of modern alternative featuring some of the best deep sky objects.

The Collinder Catalog: Per Collinder published this listing, featuring 471 open clusters, in 1931.

The Gum Catalog: The Australian astronomer Colin Gum published this listing of eighty-four emission nebulae in 1955.

The Hickson Compact Group: Paul Hickson published this listing of galaxy groups in 1982. One popular deep sky amateur target on the list is HCG 92, otherwise known as Stephan's Quintet, which is located in the constellation Pegasus.

The Index Catalog (IC): This supplement to the NGC catalog was completed in 1908 and extended the NGC listing to an additional 5,386 objects. The IC list of deep sky objects was the result of astrophotography coming into vogue at the end of the nineteenth and early twentieth century.

The Melotte Catalog: The astronomer Philibert Melotte published this supplementary list of deep sky objects in 1915. One famous Melotte object for amateurs that Messier also missed is the Coma Star Cluster Me 111 in the constellation Coma Berenices.

The Terzan List of Globular Clusters: A supplementary catalog of eleven faint globular clusters.

The Trumpler Open Cluster Catalog: Robert Julius Trumpler published this comprehensive list of thirty-seven open clusters dotting the night sky.[8]

The Uppsala General Catalog of Galaxies (UGC): Published in 1973, the UGC is a thorough survey of northern hemisphere galaxies down to declination -2 degrees, 30 arcminutes, and contains 12,921 galaxies down to a limiting magnitude of +14.5.

Note that many of these catalogs contain considerable overlap. For example, the famous Orion Nebula (Messier 42) also finds its way into the NGC catalog as NGC 1976, and is also referred to as Sharpless 281. Some clubs and organizations present certificates for observers who have made their way through these respective catalogs of objects. I like to simply challenge myself to see one new thing on a given night of observing, realizing I'll never see all the wonders the Universe has to offer in a short human life span.

THERE'S A UNIVERSE OF DEEP SKY OBJECTS

out there for the determined amateur astronomer with access to a huge light bucket telescope and a dark sky.

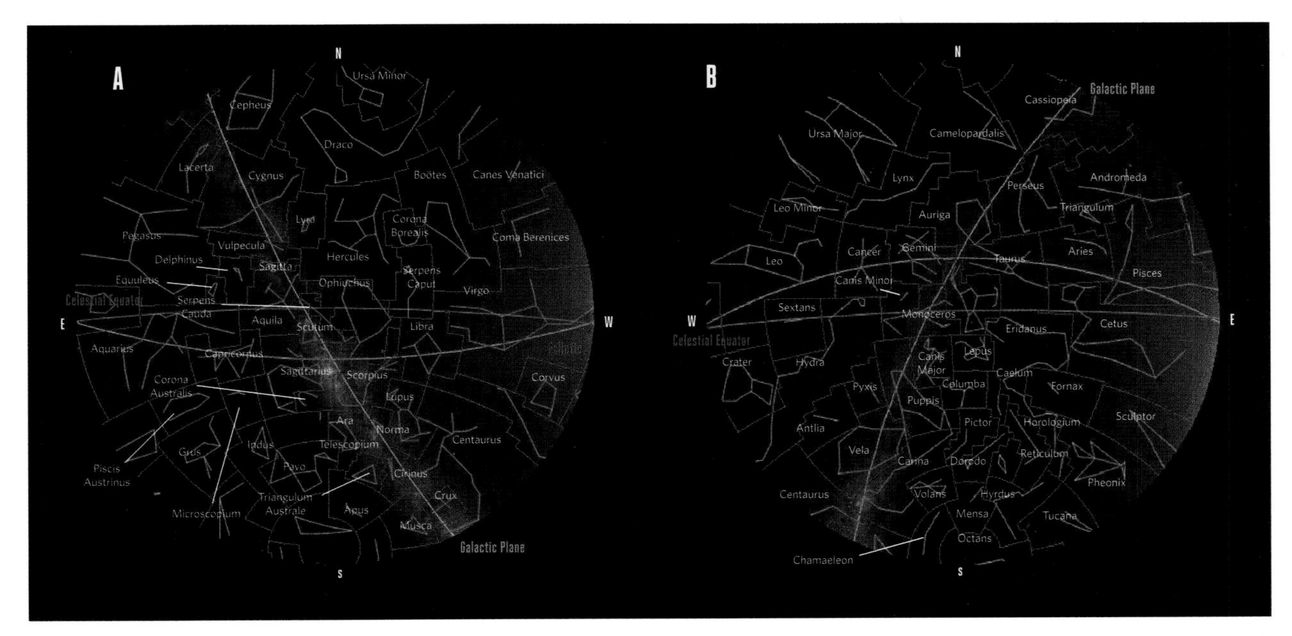

Overview charts of the two celestial hemispheres, northern (A) and southern (B), complete with constellation labels. The individual charts are centered on the respective celestial north and south poles: the blue circle is the celestial equator, the orange curve is the path of the ecliptic plane, and the brown curve is the plane of the Milky Way Galaxy. Please see larger version on page 214.

CONSIDERING CONSTELLATIONS

If we imagine the vault sky as a spherical globe overhead, then the **constellations** are the imaginary boundaries we use to section it off into regions. The sky is grouped off into eighty-eight constellations, with block-shaped angular borders following lines of right ascension and declination. The northern hemisphere constellations date back to Ptolemy's time and antiquity, though lesser variations have come and gone. For example, the January Quadrantid meteors take their name from the now-defunct constellation of Quadrans Muralis (the Mural Quadrant), which is now divvied up between the modern constellations of Draco, Hercules, and Boötes. The Babylonians, Chinese, and other cultures all saw different shapes in the heavens . . . feel free to come up with your own.

Johann Bayer's *Uranometria* star atlas, published in 1603, set the stage for the modern layout of the northern hemisphere constellations, right on the cusp of the telescopic era. The celestial cartographer Petrus Plancius first organized the southern hemisphere constellations in the late sixteenth century. The French astronomer Nicolas de Lacaille finished the job in 1763, adding seventeen more constellations to the southern hemisphere sky.[9] See appendix (page 214), Constellations of the Sky.

The Babylonians, Chinese, and other cultures all saw different shapes in the heavens . . .

The Helix Nebula (NGC 7293) is located about 700 light-years away in the astronomical constellation of Aquarius, the Water Bearer.

TYPES OF DEEP SKY OBJECTS

Besides the Moon, planets, and stars, the sky actually contains a surprisingly small array of subcategories of deep sky objects. Inhabiting the realm of astronomical objects beyond our solar system informally is what's referred to as deep space.

DIFFUSE NEBULAE

These irregular blotches of gas glow from either **reflection** or **emission**, and are often the birthplace of new stars just starting to push back their cocoons of gas, revealing themselves to the Universe. Two great examples of this are the Orion Nebula (M42) and the Carina Nebula (NGC 3372). Irregular nebulae are also sometimes the dead shrouds of exploded supernovae stars seeding their elements back into the Universe, such as the famous Crab Nebula (M1) in the constellation Taurus.

DARK NEBULAE

Notice dark splotches along the plane of the Milky Way? These are intervening patches of gas and dust known as Dark Nebulae. One of the most famous examples is the coal sack in the constellation of the Southern Cross.

PLANETARY NEBULAE

The very name "planetary nebula" is more visually than physically descriptive, and belies their true nature. Though they may look like tiny incorporeal dots, planetary nebulae have nothing to do with planets. Nicknames such as the Saturn Nebula and the Ghost of Jupiter simply refer to their appearance at the eyepiece. A planetary nebula is a star such as our Sun in the final death throes of its life, puffing smoke rings of gas and dust out at the cosmos. Each one is unique, and seems to depend on the magnetic field configuration of the spinning, dying star within. It's sobering to think that our Sun will one day appear as a tiny dull ring of gas, perhaps for the amusement of an alien star party hundreds of light-years away and billions of years from now.

GLOBULAR CLUSTERS

Can you say "GLOBE-ular?" Then you too can properly pronounce this tiny knotted class of star clusters. At the eyepiece, globular clusters look as if someone scattered stars like powdered sugar across the sky. These are also some of the most metal-poor and ancient structures in the Universe, and our Milky Way Galaxy possesses some estimated 150 globular clusters,[10] with the intervening bulk of our home galaxy probably hiding many more beyond. It's a bit frustrating that we can't hop out of the plane of our Milky Way Galaxy several thousand light-years out for a better view.

The early **Population II stars** seen in globular clusters are so old, they were once part of a cosmological controversy until just a few decades ago: How could stars in the Universe be older than the Universe itself? After all, you can't be older than your own parents, right? Today, we know, thanks to Planck data, that the refined age of the Universe is 13.7 billion years +/- 120 million years.[11] Discrepancy resolved.

OPEN CLUSTERS

Think globulars, minus the tight, unresolved core of stars. Though some open clusters are quite spread out and simply look like a slight over-density of stars at the eyepiece, some—like the famous Double Cluster in Perseus and M35 in the constellation Gemini—are quite dramatic. A few, such as the Pleiades (M45) in Taurus and the Beehive Cluster (M44) in Cancer are visible to the naked eye and spill over the edges of even a low-power field of view. The V-shaped Hyades in the head of Taurus the Bull is one of the closest deep sky objects in the sky, at 153 light-years away. As with globular clusters, you can imagine what the glorious night sky scene from a planet orbiting a star inside such a cluster, with several hundred stars

Our solar system's home address: in the Orion Spur of the Perseus Arm of the Milky Way Galaxy.

outshining Sirius in brightness. Looking at M44, M45, and the Hyades, we see not only a good study in contrast from a distant to a nearer cluster, but how open clusters evolve and disperse over time.

Our Sun and its forgotten siblings began their halcyon days inside some unknown open star cluster billions of years ago, and are now strewn around the Milky Way Galaxy and forever separated from one another.

GALAXIES

It's hard to imagine today—less than a century ago, our Milky Way Galaxy was considered the sum expanse of the Universe. You can certainly see how those nineteenth-century astronomers got there . . . filling a cylindrical-shaped volume of space 100,000 light-years wide by 1,000 light-years thick, the Milky Way certainly *seems* more than big enough. You can even still find old star charts from the late nineteenth century and the early twentieth century that list objects such as the Pinwheel and Andromeda Nebula!

NOTES ABOUT NOVAE AND SUPERNOVAE

We often think of the deep sky as unchanging . . . but there's one altering aspect to watch for: stars that occasionally go **nova** or even **supernova**. Johannes Kepler witnessed the last naked eye supernova in our galaxy in 1604; you could say we're due.[12] More common are supernovae visible in other galaxies, such as SN 2014J, which occurred in January 2014. These usually don't top +10th magnitude, however, and require a good-size telescope to see. Amateur astronomers still occasionally discover supernovae in other galaxies, and these can change the face of a galaxy for several weeks as the supernova briefly outshines all of the host galaxy's other stars.

More common are novae or stars flaring up in our own galaxy. These occur on average every few years. Nova Delphini 2013 shone at magnitude +4.3, briefly changing the face of the diamond-shaped asterism. The American Association of Variable Star Observers (AAVSO) is the central clearing house for variable stars and variable star astronomy, and periodically issues alerts for supernovae and novae.

ACTIVITY: MESSIER MARATHONING

Did you know that you can actually see all 110 of the deep sky objects in Messier's famed catalog . . . in one night? The best time to complete this feat is near the northward March equinox around the dark weeks of the New Moon. This feat of visual athletics works because a majority of Messier's objects sit near the galactic plane, and only a few key challenge objects can trip up a prospective Messier marathoner's stride.

For example, you'll want to nab M77 and M74 immediately after dusk, or the race is over right out the gate. From there, you'll want to work over the sky eastward in groups, going through the Andromeda Galaxy (M31) and nearby objects over through Orion and M42 and its neighbors. From there, you can lengthen your pace and slow down a bit as the night unfolds. Then it's on to a more leisurely stroll through Leo and the Virgo and Coma Berenices galaxy groups, then eastward to a jaunt across the plane of the Milky Way Galaxy from Sagittarius through Cygnus the Swan, flying high near local dawn. Pick up the pace now, as M72 and M73 in the constellation Aquarius and the globular cluster M30 in Capricornus are the toughest objects of all, visible low to the southeast just before the brightening dawn overtakes them.

Can you pick 'em off with binoculars? How many Messier objects can you see with the naked eye from your favorite dark sky site? How about conducting a photographic Messier marathon, where you try to image all the objects in the catalog in one night? Whether you leisurely pick off Messier's catalog over a night, a year, or a lifetime, a tour through the list is a great way to familiarize yourself with the wonders of the night sky and whet your appetite for going deeper into the NGC and other catalogs.

Good luck, and Happy Messier marathoning!

There are lots of great resources for exploring Messier's catalog online, including this one via SEDS.org: http://www.messier.seds.org/

The best Messier marathon primary and secondary weekends for the next five years are shown in the table on the next page.

MESSIER MARATHON DATES FOR THE NEXT FIVE YEARS

Year	New Moon	Primary	Secondary
2019	March 6	March 30	March 9
2020	March 24	March 21	March 28
2021	March 13	March 13	None
2022	March 2	April 2	March 5
2023	March 21	March 18	March 25

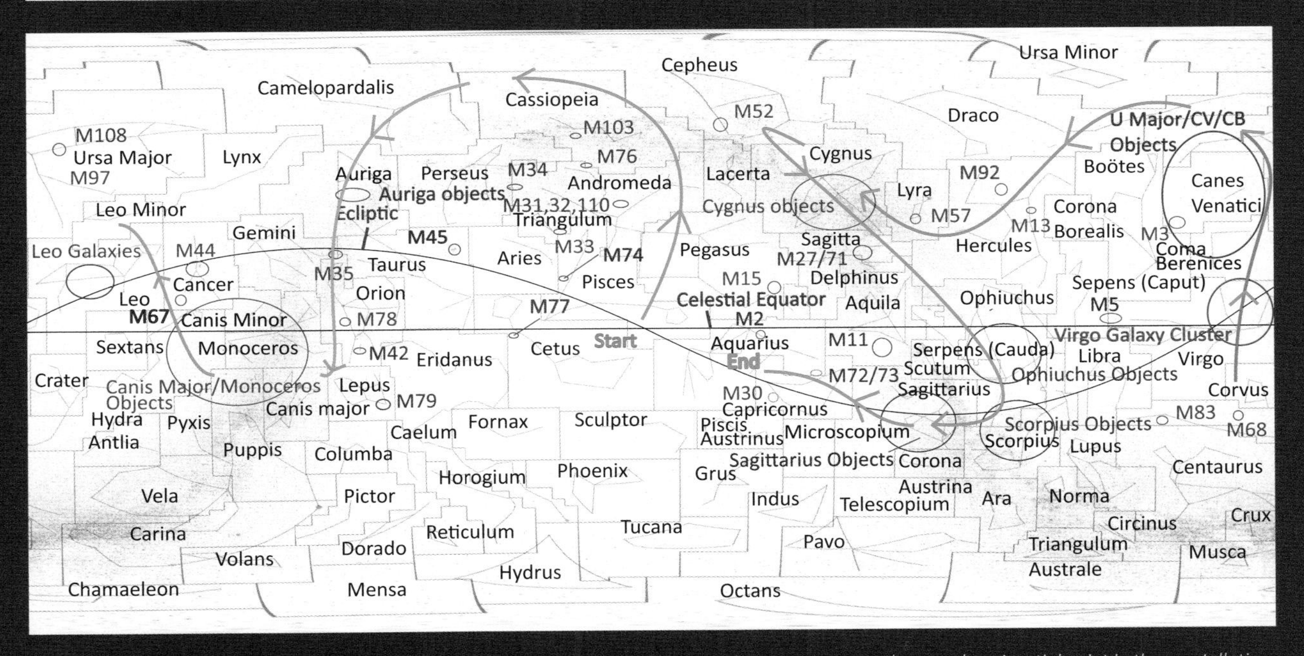

Chutes and Ladders, *celestial style: a suggested path for a springtime Messier marathon, starting near the vernal equinoctial point in the constellation Pisces in the evening and ending in the early morning in the nearby constellation of Aquarius.*

THE SKY FROM SEASON TO SEASON

A tour of the night sky through the seasons.

"Space is big. Really big. You just won't believe how vastly, hugely, mind-bogglingly big it is. I mean, you may think it's a long way down the road to the chemist's, but that's just peanuts to space."

— *Douglas Adams*, The Hitchhiker's Guide to the Galaxy

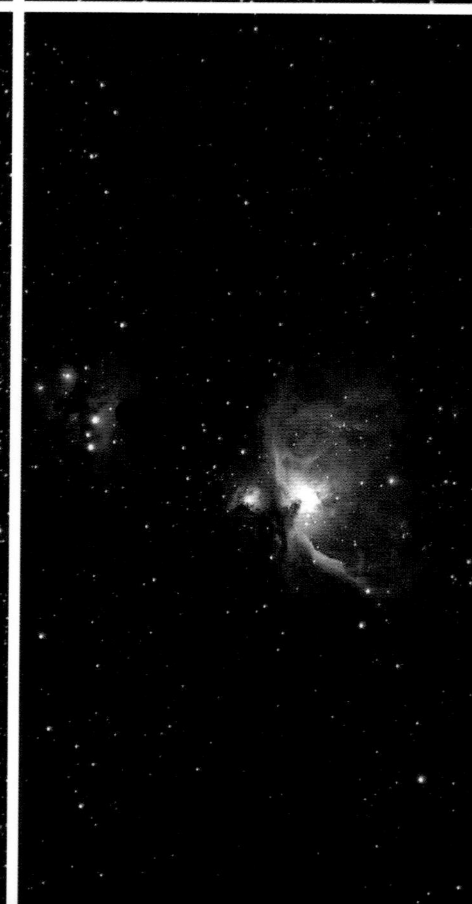

Study the deep sky long enough, and its wonders become like greeting the return of old friends each season. In this chapter, we'll show you how to find your way around the night sky from season to season, the parade of star party favorites, and a few secret weapons to show off when everyone else's telescope tube is pointed Saturn-ward.

If you go out and look at the sky every night at the same time, you'll see nearly the same sky view from one night to the next . . . almost. The vault sky slowly shifts almost one full degree, or just under 4 minutes of arc in right ascension, a day. This motion is, of course, an illusion, reflecting Earth's motion once a year around the Sun. To put it another way, we see the sky present overhead two new hours of right ascension rising to the east (as the constellations move *westward* from the observer's perspective) per calendar month. Your best views are always along the meridian (the line that bisects the vault of the sky from north to south) and the zenith, where you're looking through the thinnest layer of the atmosphere, straight up overhead.[1] It's helpful to think of the sky scene from month to month as when the hour of the particular constellation of the zodiac is perched along the north–south meridian line of the sky. You can, of course, select whatever viewing hour you want, changing the sky scene accordingly.

For the purpose of this guide, we'll look at the sky scene for the northern hemisphere from 40 degrees north latitude at 10 p.m. Standard Time at the beginning of the four astronomical seasons (the March and September equinoxes, and the June and December solstices), including finding your way around the sky for the particular season, and a selection of the very best deep sky objects.

The sky shifts 360 degrees/4 seasons = 90 degrees, or 6 hours of right ascension from one season to the next.[2]

Note that, unless a target is faint or especially difficult to find, we simply list its position in right ascension (R.A.) and declination (Dec.) rounded off to the nearest half-arcminute, more than accurate enough to place the object in the eyepiece at medium magnification. The equatorial R.A./Dec. system is the very best way to reference the fixed location of deep sky objects, as it barely changes over the span of several decades and is independent of the seasons and the local horizon. In astronomer-speak, (') is shorthand for "arcminute," and (") is short for "arcsecond."

ACTIVITY

- Nabbing southern sky objects . . . from up north (page 100)

THE CONSTELLATIONS URSA MAJOR AND LEO DOMINATE THE MARCH SKY . . .

THE MARCH SKY

THE NIGHT SKY AT CANCER HOUR

The springtime evening night sky for the northern hemisphere is a view in transition. The Orion Spur (within which our Sun boasts its galactic address)[3] and the Perseus Arm of the Milky Way Galaxy are sinking into the dusk sky to the west, along with the winter hexagon of bright stars (see The December Sky, page 92). The constellations Ursa Major and Leo dominate the March sky, along with the promise of the Virgo and Coma Berenices clusters of galaxies in late spring and early summer. The plane of the Milky Way and the Perseus Arm looking out away from the galactic core bracket these constellations as they set to the west, while the September southward equinoctial point that the Sun will occupy in 6 months rises in the east.

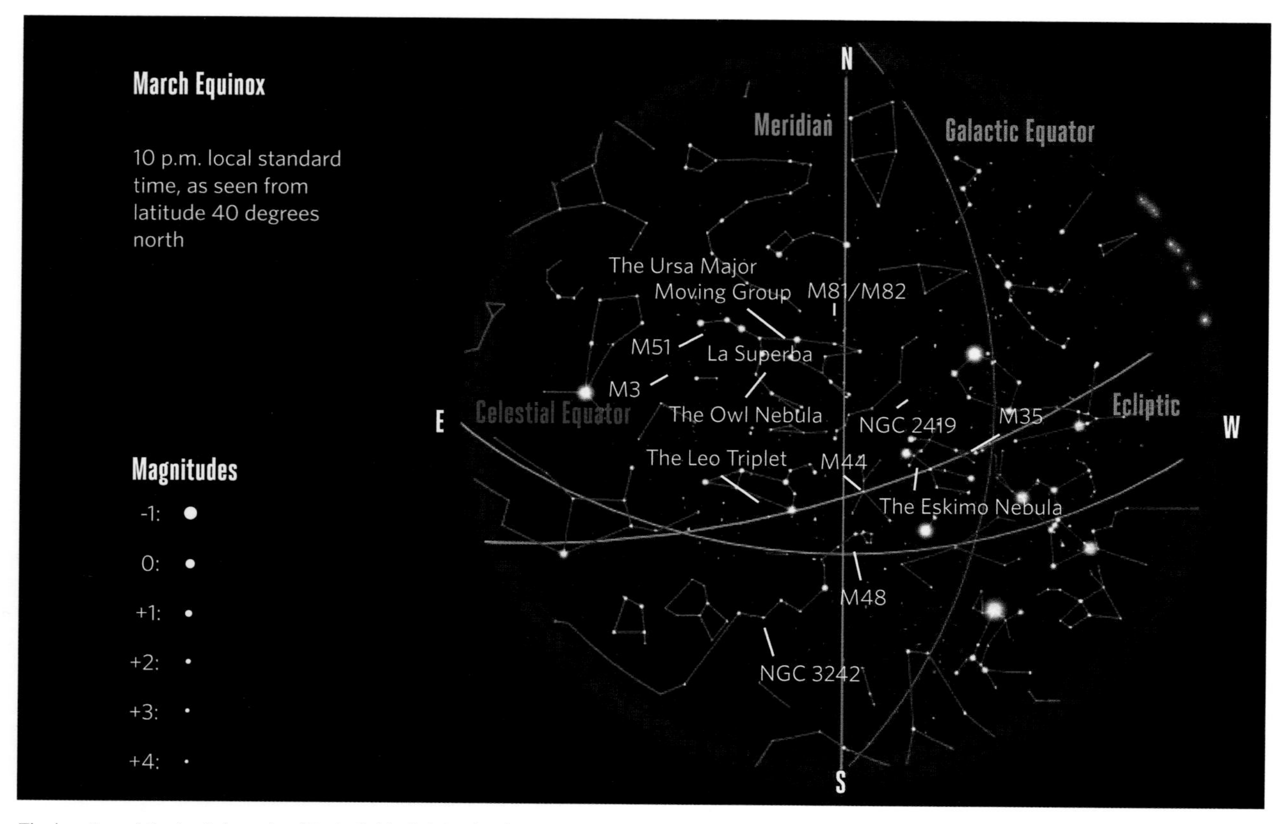

The location of the best deep sky objects (white) rising for the spring season, as seen from latitude 40 degrees north around 10 p.m. local standard time.

BEST OBJECTS

The Leo Triplet: A fine grouping of 9th–10th magnitude spiral galaxies (M65, M66, and NGC 3628). Located 35 million light-years away in the constellation Leo the Lion, the Leo Triplet spans a 45-arcminute field of view, forming a sideways "\:" face staring back at the observer across time and space.

Coordinates: (centered on M66) R.A. 11H 20′ Dec. +12° 59.5′

The Owl Nebula (Messier 97): Located 2 degrees southeast of the +2.3 magnitude star Merak (Beta Ursa Majoris) in the corner of the bowl of the Big Dipper asterism, M97 is one of the tougher objects in Messier's catalog. This planetary nebula shines at magnitude +10, but the trouble is, that brightness is smeared out over an apparent diameter of over 3 arcminutes. Under dark skies, you might just spy the two dark patches that give M97 the appearance of an owl's face staring back at you from 2,030 light-years away. Also, keep an eye out for the +10.7 magnitude galaxy M108 nearby.

Coordinates: R.A. 11H 15′ Dec. +55° 01′

M35: I love showing off this gem of an open cluster in the foot of the constellation Gemini the Twins. Spanning the diameter of a Full Moon, this structured cluster shines at +5th magnitude and is visible to the naked eye from a dark sky site. M35 is an easy catch with binoculars. Sitting very close to the ecliptic plane at an estimated 2,800 light-years away, the Sun crosses very near M35 on the June solstice, and planets often transit across its face.

Coordinates: R.A. 6H 09′ Dec. +24° 21′

M81 and M82: I never miss an opportunity to nab this fine pair of galaxies in Ursa Major. If I can easily snap up these two 7th and 8th magnitude galaxies at low power, then I know the seeing is steady and it's a good night for deep sky observing. About a Full Moon's diameter (30′ [9 m]) apart and 12 million light-years away, these opposing galaxies offer a good study in contrast with M81 (sometimes called Bode's Nebula) presenting a flat-on face, and the Cigar Galaxy M82 showing off a nearly edge-on view. These two galaxies are monsters, both bigger than our own Milky Way. Though they're fine targets, they also lie in a rather star-poor region of the northern sky about 8 degrees west of the +3.8 magnitude star Lambda Draconis. This paucity of foreground stars is why NASA chose a region 15 degrees east of the M82/M81 pair in northern Ursa Major for the 1995 Hubble Deep Field.

Coordinates (centered on M81): R.A. 9H 56′ Dec. +69° 04′

The Eskimo Nebula NGC 2392: This one simply *looks* astounding in deep sky images, but may be a bit underwhelming to spot in person. The Eskimo Nebula is a +9.5 magnitude planetary nebula, about 45 arcseconds across. Crank up the magnification, and you just might see hints of the fuzzy-hooded face of the Eskimo. For our money, we think the Eskimo Nebula is easier to pick out than M97, which vanishes with the least bit of loss of sky transparency. Still, we can forgive Messier for passing over its more fuzzy, starlike veneer with his eighteenth-century optics. The Eskimo Nebula is a short 2-degree hop southeast of Delta Geminorum.

Coordinates: R.A. 7H 29′ Dec. +20° 55′

M48 (NGC 2548): The forty-eighth object listed in Messier's catalog is also one of its great mysteries, as no bright deep sky object exists at the precise coordinates given. Instead, Messier 48 is usually ascribed to a nearby +5.5 magnitude open star cluster NGC 2548 in Hydra near its border with Monoceros. Was Messier's entry a transcription error, or perhaps a spurious and as yet unrecovered comet? The Open Cluster M48 (NGC 2548) spans about 54 arcminutes and is about 1,500 light-years away.

Coordinates: R.A. 8H 14′ Dec. -5° 45′

NGC 2419: One of the most distant globular clusters known at 300,000 light-years away, NGC 2419 is often dubbed the Intergalactic Wanderer, though it is, in fact, bound to our galaxy. NGC 2419 takes an amazing 3 billion years to make one orbit. Imagine the view on its fringes, looking back at the magnificent Milky Way Galaxy! NGC 2419 spans about 6 arcminutes and shines at magnitude +9. The cluster sits 7 degrees north of the bright star Castor in the otherwise nondescript constellation of the Lynx.

Coordinates: R.A. 7H 38′ Dec. +38° 53′

M51: This fine face-on swirling galaxy was also one of the first nebulae to betray its true spiral structure. Known as the Whirlpool Galaxy, M51 measures 11 by 7 arcminutes across and shines at +8th magnitude. Look closely, and you might spy a lobe to one side, where M51 is interacting with the dwarf galaxy NGC 5195. This pair looks amazing in large apertures under dark skies, revealing dark spiral dust lanes alternating with bright areas of star formation. The pair is 25 million light-years away, and M51 is in the constellation Canes Venatici 3 degrees southwest of the +2nd magnitude star Alkaid (Eta Ursae Majoris).

Coordinates: R.A. 13H 30′ Dec. +47° 12′

M3: One of the best deep sky objects in the northern hemisphere sky, Messier 3 is a fine globular cluster, easily resolved into individual stars even at low power. In fact, I can usually sweep up +6th magnitude M3 simply by scanning the field with binoculars. M3 lies in a star-poor region about a third of the way from the star Arcturus in Boötes to Alkaid in Ursa Major, though it's actually located in the constellation Canes Venatici. M3 is about 34,000 light-years away.

Coordinates: R.A. 13H 42′ Dec. +28° 23′

M44: Often referred to as the Beehive Cluster, this open cluster is a naked eye patch in the heart of Cancer the Crab, and is a centerpiece for the constellation. Also known as Praesepe (Latin for The Manger), M44 was known as a nebulous patch since at least ancient Greek times. Even at low power, the hundreds of stars in the Beehive Cluster live up to their name, spilling over and swarming across a degree and a half field of view. It's not uncommon for a bright planet to transit in front of the +3rd magnitude open cluster, making for a photogenic view.

Coordinates: R.A. 8H 40′ Dec. +19° 59′

Locating "La Superba" on the Canes Venatici-Ursa Major border.

NGC 3242: Messier probably missed this fine planetary nebula due to its location, deep in the southern realms of the snaking constellation Hydra. Often called the Ghost of Jupiter due to its spectral 25-arcsecond diameter blue-gray disk, NGC 3242 is 1,400 light-years away. This planetary nebula sits just under 2 degrees south of the +3.8 magnitude star Mu Hydrae.

Coordinates: R.A. 10H 25′ Dec. -18° 38′

The Ursa Major Moving Group: Finally, I'd like you to step back from the eyepiece and to turn your attention to the big picture the springtime evening sky paints overhead, post dusk. If folks only know one or two constellations, it's generally Orion the Hunter (now setting low to the west at dusk), or the Big Dipper (actually an asterism formed from the brightest stars in the constellation Ursa Major, the Great Bear). In 1869, astronomer Richard Proctor was making a survey of the proper motions of bright stars in the northern hemisphere, when he came upon a revelation: Most of the bright stars in Ursa Major and the adjacent constellation Canes Venatici are headed in the direction of the constellation Sagittarius. Today, we know that this Moving Group and the related Ursa Major Stream of stars claims alumni in several scattered constellations, including Auriga, Lepus, and Orion. This also represents the closest loose cluster of stars near the Earth, at only 80 light-years away.

Curiously, the Sun is not a member of this younger stream of stars, but is instead just slipping past the Ursa Major Moving Group, like a cosmic renegade.

Challenge object: La Superba: Sure, the Universe is a pretty gray-white place to the naked eye . . . but one class of objects that buck this trend are known as **carbon stars**. These ruby gems can startle viewers at the eyepiece, their ruddy hues a product of ancient red giant stars fusing carbon at the end of their lives. Such is the fate of our Sun, giving one last encore performance as a cherry-red beacon before retiring to a degenerate (not commentary on its moral state) **white dwarf**. A fine example of a carbon star is the variable Y Canum Venaticorum in the constellation Canes Venatici, near its border with Ursa Major. Nineteenth-century Jesuit astronomer Father Angelo Secchi christened this star "La Superba" (The Magnificent).[4] La Superba varies from magnitude +4.9 to +7.3, a change of 2.4 magnitudes in brightness over roughly every 160 days.

Coordinates: R.A. 12H 45′ Dec. +45° 26.5′

THE JUNE SKY

THE SKY AT LIBRA HOUR

The beginning of astronomical summer in the northern hemisphere sees the view overhead rotate westward, past the distant cluster of galaxies boiling out of the bowl of Virgo, and into the streamers of Berenices' Hair and the constellation Coma Berenices. While this distant view passes the zenith to the west, the magnificent plane of the Milky Way Galaxy rises to the east, bringing with it the promise of the star-rich fields of Sagittarius and Scorpius due south on mid- to late-summer evenings. The Summer Triangle asterism, made up of the bright stars Vega, Deneb, and Altair, is the hallmark of the summer sky. Look toward the constellation Hercules southwest of the star Vega, and you're staring headlong into the **Apex of the Sun's Way**, the direction in which our Solar System is hurdling around our galaxy at 137 miles (220 km) per second.

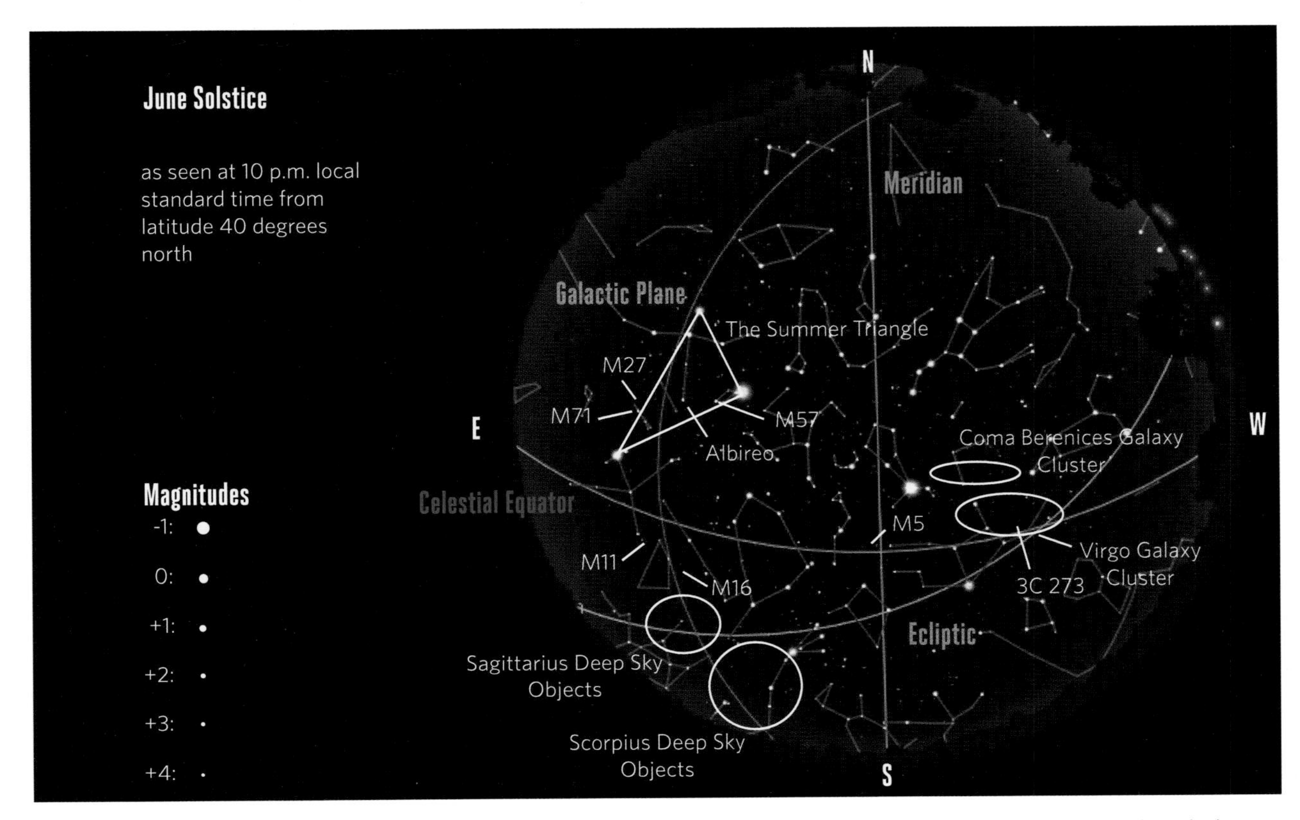

The location of the best deep sky objects (white) rising for the summer season, as seen from latitude 40 degrees north around 10 p.m. local standard time.

BEST OBJECTS

The Virgo and Coma Berenices clusters of galaxies: One of the great highlights of the late spring/early summer sky is the cascade of galaxies across the constellations Virgo and Coma Berenices. Plop your view down in the Bowl of Virgo at low power, and you'll probably stumble across a random galaxy after sweeping around for a minute or two. Top objects in Virgo to watch for (with magnitudes) are: M60 (magnitude +9.5), M49 (+9.4), M86 (+9.8), M87 (+9.6), and M84 (+10.1). This tight knot of bright galaxies is centered on R.A. 12H 43′ Dec. +11° 33′ and extends southeastward.

The top objects in the constellation Coma Berenices (with magnitudes) are M100 (magnitude +10.1), M85 (+10.0), and M64 (+9.4). Also, don't miss the fine +8th magnitude globular cluster M53 at R.A. 13H 13′ Dec. +18° 10′.

M57: I once heard of an apt description of the famous Ring Nebula in the constellation Lyra at a star party as the "ghost of a doughnut." An easy find about midway between Beta and Gamma Lyrae, M57 is 4 arcminutes across and about 2,300 light-years away. A good challenge for large light bucket owners is to try and nab the +15th magnitude central white dwarf star at the center of the nebula.

Coordinates: R.A. 18H 54′ Dec. +33° 02′

M13: A mandatory summer deep sky object and one of the best globular clusters in the northern sky, +5.8 magnitude M13 in Hercules never fails to impress. M13 is located 2.5 degrees south of the 3.5 magnitude star Eta Herculis, bordering on the edge of the Keystone asterism.

Coordinates: R.A. 16H 42′ Dec. +36° 28′

M16: Remember the Hubble Space Telescope team's dramatic Pillars of Creation images released back in 1995?[5] Those tendrils of glowing gas are part of +6th magnitude M16 in the constellation Serpens the Serpent, known as the Eagle Nebula. The Pillars image actually spans a very small portion of the nebula, which is located an estimated 7,000 light-years away.

Coordinates: R.A. 18H 19′ Dec. -13° 49′

M27: Not all planetary nebulae are ring-shaped. A fine example is the apple core–shaped M27 in the small constellation Vulpecula the Fox, sometimes referred to as the Dumbbell Nebula. Eight by six arcminutes in apparent size across and shining at magnitude +7.5, M27 holds up even under light-polluted skies. M27 is a fine example of a planetary nebula, seen from an oblique side angle.

Coordinates: R.A. 19H 59.5′ Dec. +22° 43′

M5: Another fine, though often overlooked, +6.7 magnitude globular cluster in the constellation Serpens. Perhaps M5 is often passed up due to the wealth of deep sky targets along the summertime galactic plane. M5 is located about 25,000 light-years away, about 1,500 light-years less than the distance of core of the Milky Way Galaxy from Earth.

Coordinates: R.A. 15H 18.5′ Dec. +2° 05′

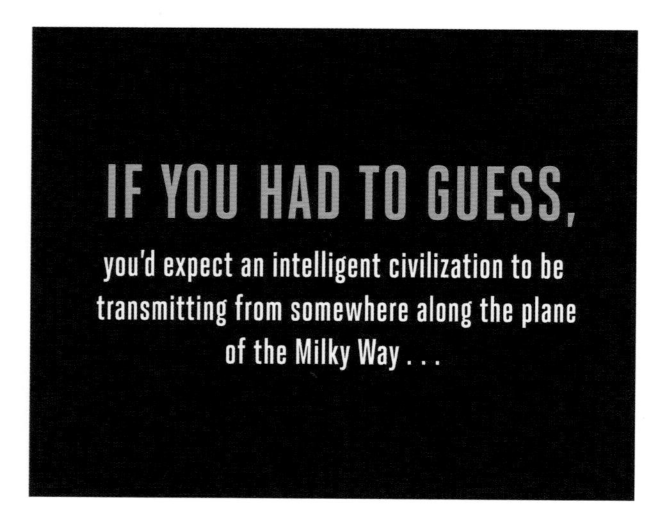

M11: Located just 4 degrees southwest of the +3.4 magnitude star Lambda Aquilae, this +6.3 magnitude open cluster in the constellation Aquila the Eagle has the evocative nickname of the Wild Duck Cluster. This is a great object to sweep up with binoculars on a summer evening. M11 is located 6,200 light-years away.

Coordinates: R.A. 18H 51′ Dec. -6° 16′

M71: The tiny constellation of Sagitta the Arrow houses a big delight: the +6.1 magnitude globular cluster M71. Unlike most globular clusters, M71 does not display a dense core, resulting in a bit of an identity crisis as an open versus globular cluster. Studies in the 1970s confirmed the low metallicity of the stellar population in the cluster, a win for the globular camp. M71 is an easy find, about midway between Gamma and Delta Sagittae.

Coordinates: R.A. 19H 54′ Dec. +18° 47′

Sagittarius objects: The southernmost constellation of the zodiac is Sagittarius the Archer, but it's always the most recognizable from its angular-shaped, Teapot asterism of stars. You're now looking toward the shrouded core of our Milky Way Galaxy, just above the tip of the spout. From up north, Sagittarius just peeks above the southern horizon on summer evenings; from the equator southward, it is a glorious object, straddling the sky high overhead. Travel south of the equator, and you'll realize you've never *really* seen the Milky Way before.

The field is also littered with deep sky objects, like clouds of steam puffing from the teapot's spout. It's a rewarding view, to simply sweep the field with binoculars. Some key highlights are listed in the table on the next page.

TOP DEEP SKY OBJECTS IN THE CONSTELLATION SAGITTARIUS					
Name	Type	Magnitude	Size	R.A.	Declination
M23	Open Cluster	6.9	27′	17H 57′	-19° 01′
M18	Open Cluster	7.5	9′	18H 20′	-17° 08′
M25	Open Cluster	4.6	32′	18H 32′	-19° 07′
M21	Open Cluster	6.5	13′	18H 05′	-22° 29′
M20	Emission Nebula	6.3	28′	18H 02′	-23° 02′
M8	Emission Nebula	6	90′ x 45′	18H 04′	-24° 23′
M17	Emission Nebula	6	11′	18H 20′	-16° 11′
M28	Globular Cluster	7.7	11′	18H 24.5′	-24° 52′
M22	Globular Cluster	5.1	32′	18H 36′	-23° 54′
M54	Globular Cluster	8.4	12′	18H 55′	-30° 29′
M70	Globular Cluster	9.1	8′	18H 43′	-32° 17.5′
M69	Globular Cluster	8.3	10′	18H 31′	-32° 20′
M55	Globular Cluster	7.4	19′	19H 40′	-30° 58′
M75	Globular Cluster	9.2	7′	20H 6′	-21° 55′

Another area of note in Sagittarius is the patch of sky very near the +3.3 magnitude star Tau Sagittarii at the bottom knurl of the teapot handle. This spot was where Ohio State University's Big Ear radio receiver picked up the mysterious Wow! signal on August 15, 1977.[6] The region has been the subject of intense scrutiny since, with nary a peep. If you had to guess, you'd expect an intelligent civilization to be transmitting from somewhere along the plane of the Milky Way, where there's simply more stars clustered along our line of sight. There's not much to see around Tau Sagittarii, but it's interesting to contemplate just what the signal was.

Scorpius objects: I love the constellation Scorpius the Scorpion; unlike most of the astronomical constructs in the sky, its snaking form looks exactly like what it refers to. Orange Antares (literally the anti-Ares or anti-Mars) is the crown jewel of the constellation. A wealth of deep sky objects spill over from adjacent Sagittarius into Scorpius as well.

TOP DEEP SKY OBJECTS IN THE CONSTELLATION SCORPIUS					
Name	Type	Magnitude	Size	R.A.	Declination
M7	Open Cluster	3.3	80′	17H 54′	-34° 47.5′
M6	Open Cluster	4.2	25′	17H 40′	-32° 13′
M4	Globular Cluster	5.9	26′	16H 23.5′	-26° 31.5′
M80	Globular Cluster	7.9	10′	16H 17′	-22° 58.5′
NGC6231	Open Cluster	2.8	15′	16H 54′	-41° 50′

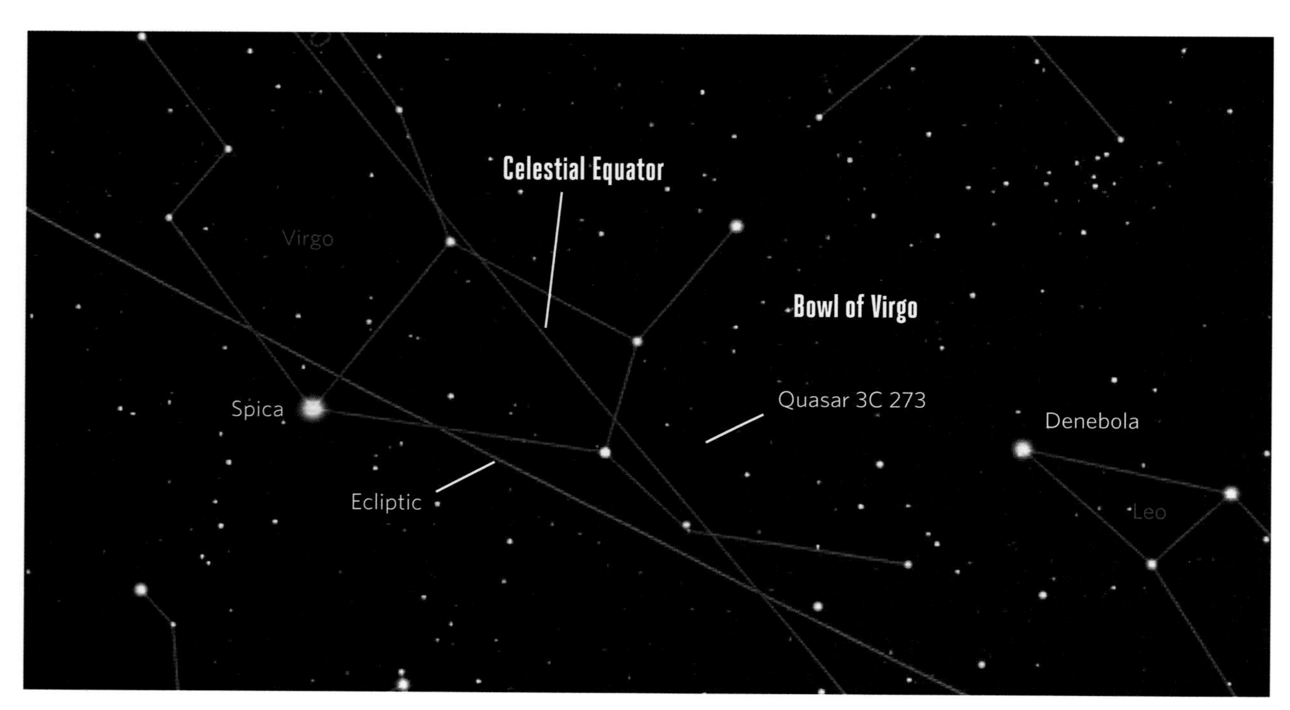

A widefield finder chart for the quasar 3C 273 in the Bowl of Virgo asterism, 2.4 billion light-years distant. For a fine field finder chart, check out the AAVSO: https://www.aavso.org/vsots_3c27

Quasar challenge: Looking for something really far out? In 1963, the redshift measurement of an otherwise nondescript +12.9 magnitude object related to a radio source in the constellation Virgo revealed an amazing surprise: though faint, this point of light dubbed 3C 273 is located 2.4 billion light-years away and is actually an intrinsically luminous object. 3C 273 was the very first **quasar** (short for "quasi-stellar object") ever discovered. 3C 273 is also in range of a decent-size backyard telescope under dark skies. Its coordinates are R.A. 12H 29′ 07″ Dec. +2 03′ 09″, and though it may not look like much, part of the thrill in finding 3C 273 is knowing what you're looking at!

You'll need a good finder chart (either from a planetarium program or an online source such as the AAVSO) to generate a chart down to about 13th magnitude to find 3C 273.

THE SEPTEMBER SKY

THE SKY AT CAPRICORNUS HOUR

Fall is my favorite season back home in New England. It lacks the most annoying features of the other three times of year: ice and cold (winter), mud (spring), and bugs (summer). From high northern latitudes, astronomy is a seasonal sport, as the perpetual daylight of summer and the brutal cold of winter conspire against backyard astronomers. Evenings in the northern hemisphere in the fall also see the Summer Triangle and the Milky Way give way to the Great Square of Pegasus, and the sparse constellations of Capricornus and Aquarius. The equinoctial point in Pisces is also rising, marking the spot the Sun will occupy at the spring equinox.

BEST OBJECTS

M31: The famous Andromeda Galaxy is one of the northern hemisphere's finest sights. M31 is a fine starter deep sky object to cut your teeth on, visually and photographically. Shining at magnitude +3.5, M31 is one of the brightest Messier objects and is barely visible to the naked eye. Sweep just a degree west of +4th magnitude Nu Andromedae to scoop out this fascinating galaxy. Though the least bit of light pollution will kill the subtle extensions of the galaxy, M31 actually extends over a 3 x 1 degree patch of sky, the size of several Full Moons. Also, keep an eye out for the brightest of M31's satellite galaxies, +8th magnitude M110 nearby.

Coordinates: R.A. 00H 43′ Dec. +41° 17′

M33: Another fall favorite, the +5.7 magnitude galaxy M33 in the constellation Triangulum is a teaser. As with many extended deep sky objects, the least bit of light pollution will cause most of Messier 33 to vanish entirely. Compounding the problem of hunting for M33 is the dearth of bright guide stars in this patch of sky. M33 is about a third of the way and offset just to the right of a line drawn from Alpha Triangulum and Beta Andromedae. Three million light-years away, M33 is only 20 percent farther away than M31.

Coordinates: R.A. 1H 34′ Dec. +30° 37′

M52: An open star cluster in the W-shaped constellation Cassiopeia, M52 is located near the border of Cepheus. M52 is one of the finest objects in Cassiopeia. An extended scattering of stars with an overall brightness of magnitude +5, M52 is about 5,000 light-years from the Earth.

Coordinates: R.A. 23H 24′ Dec. +61° 35′

M74: Shining at +9th magnitude, this face-on spiral galaxy in the constellation Pisces can also vanish from view under less than optimal skies. Covering a 10-arcminute span, M74 is just a degree east of the +3.8 magnitude star Eta Piscium. M74 has also been the site of several notable supernovae over the past century.

Coordinates: R.A. 01H 37′ Dec. +15°47′

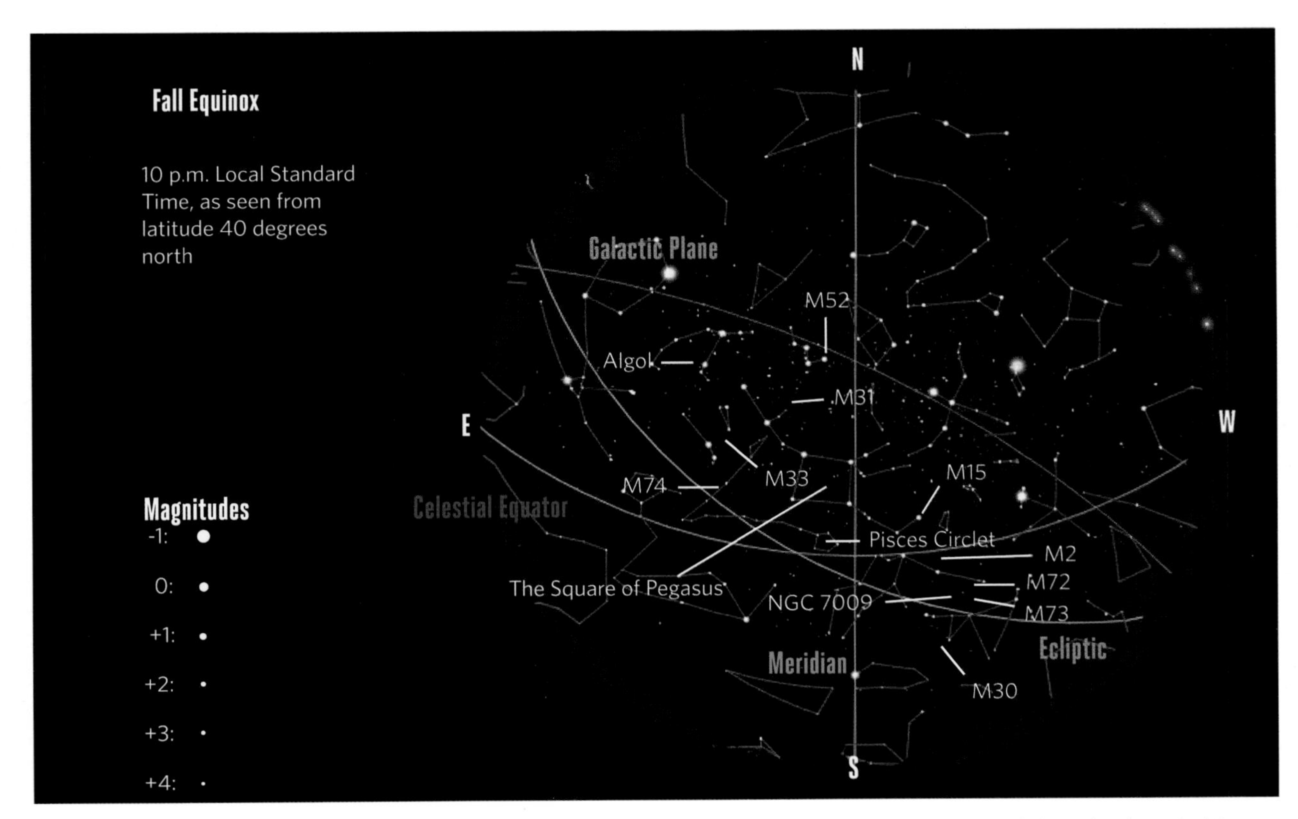

The location of the best deep sky objects (white) rising for the fall season, as seen from latitude 40 degrees north around 10 p.m. local standard time.

M15: Though Pegasus is one of the larger constellations in the sky in terms of angular size, it contains a surprising paucity of deep sky objects. The globular cluster M15 is located near the constellation's border with, ironically, the second-tiniest constellation in the sky, (next to Crux) Equuleus. Located 33,000 light-years away (about 50 percent farther than the distance to the galactic core), M15 is also one of the most ancient globular clusters known at an age of about 12 billion years old, dating from a time when the Universe was about 14 percent of its current age. Shining at magnitude +6, M15 is 4 degrees equidistant from Enif (Epsilon Pegasi) and Delta Equulei.

Coordinates: R.A. 21H 30′ Dec. +12° 10′

The Andromeda Galaxy (Messier 31) 2.5 million light-years distant. Composed using 40 x 60-second frames, focal length 388mm f/6.3, ISO 3200.

NGC 7009: As with many fine planetary nebulae, NGC 7009 in the constellation Aquarius is one that got away from Messier's catalog. Shining at magnitude +8 and about 3,000 light-years away, NGC 7009 has two opposing lobes, earning it the nickname of the Saturn Nebula. About 41 by 35 arcseconds in size, this nebula may appear a slight yellowish-green at high magnification. As with many deep sky objects in the Fall, NGC 7009 also suffers from a minor public relations crisis due to its location away from any major bright stars, making it a tough find. The Saturn Nebula is located 5 degrees southeast of the +3.7 magnitude star Epsilon Aquarii.

Coordinates: R.A. 21H 04′ Dec. -11° 22′

M2: Another curious feature of the Messier catalog is how it haphazardly skips around the northern sky, a testament to how it was compiled in a patchwork style over decades. The second object in Messier's catalog is also one of the largest globular clusters in the sky. Located five degrees north of the 2.9 magnitude star Beta Aquarii, +6th magnitude M2 would be a truly magnificent object were it not so distant, at an estimated 33,000 light-years away.

Coordinates: R.A. 21H 33.5′ Dec. -00° 49′

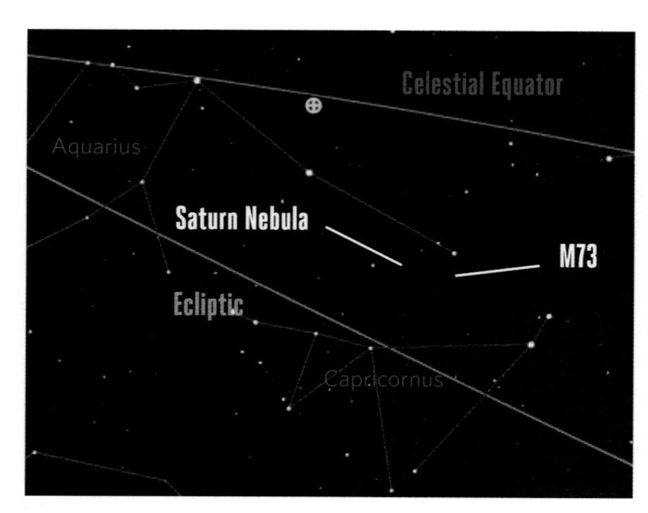

Locating the Saturn Nebula and Messier 73 on the Aquarius-Capricornus border.

M72: One of the more difficult objects on Messier's list, M72 is a remote globular cluster 55,000 light-years away in the constellation Aquarius. At +9th magnitude, this fuzz ball of a cluster resists resolution, and requires hefty magnification to resolve its diffuse core into individual stars. For this reason, many early deep sky surveys list M72 as a nebula instead of a true star cluster. M72 is a short 4-degree hop west of the Saturn Nebula. (See page 91, NGC 7009.)

Coordinates: R.A. 20H 53′ Dec. -12° 32′

M30: One of the best objects in the constellation Capricornus the Goat, +7th magnitude M30 is a 12-arcminute globular cluster located 27,000 light-years away. M30 is located about 4 degrees south of the +3.8 magnitude star Zeta Capricorni.

Coordinates: R.A. 21H 40′ Dec. -23° 11′

Astro Challenge: What was M73? The Messier catalog presents the modern-day observer with a few mysteries, not the least of which is the nature and identity of its seventy-third entry. Looking at the given location in the constellation Aquarius within a degree of M72, we find a nondescript, loose asterism containing a handful of +9th magnitude stars . . . is this a testament to how bad Messier's optics were, or did galactic tidal forces rip M73 apart sometime in the past into the very diffuse cluster we see today? The controversy surrounding the group lingered right up to the early twenty-first century, when the Max Planck Institute completed a survey in 2000 revealing that the suspect core stars are at very different distances and moving in dramatically different directions,[7] meaning they're gravitationally unbound and unrelated.

Coordinates: R.A. 20H 59′ Dec. -12° 38′

Sometimes, it seems, the Universe fooled even a seasoned observer like Messier.

THE DECEMBER SKY

THE SKY AT ARIES HOUR

Winter observers, take heart; the wintertime sky also holds some of the best sights. If you can brave the frigid air, you might also encounter very good stable seeing on a clear December evening. Once again, the plane of the Milky Way and the ecliptic rides high to the southeast at dusk, only this time, you're looking *outward* from our Solar System's home address in Orion Spur of our Milky Way Galaxy, toward the adjacent Perseus Arm. This also means you've got lots of nearby stars in this region, including Sirius, the brightest star in the sky at magnitude -1.5 only 8.6 light-years away. Sirius makes up one point of the Winter Hexagon, which also includes (counterclockwise) Rigel, Aldebaran, Capella, Castor and Pollux, and Procyon. These all circle brilliant ruddy Betelgeuse in the shoulder of Orion, located 498 light-years away. We know that Betelgeuse is at the end of its life, and will grace the skies of Earth with a brilliant naked eye supernova . . . soon. Of course, "soon" is a relative term when you're talking about astrophysical events; light from a Betelgeuse-turned supernova *could* be racing Earthward now, set to reach human retinas tonight . . . or the historic event could still be tens of thousands of years off. *But it will happen.*

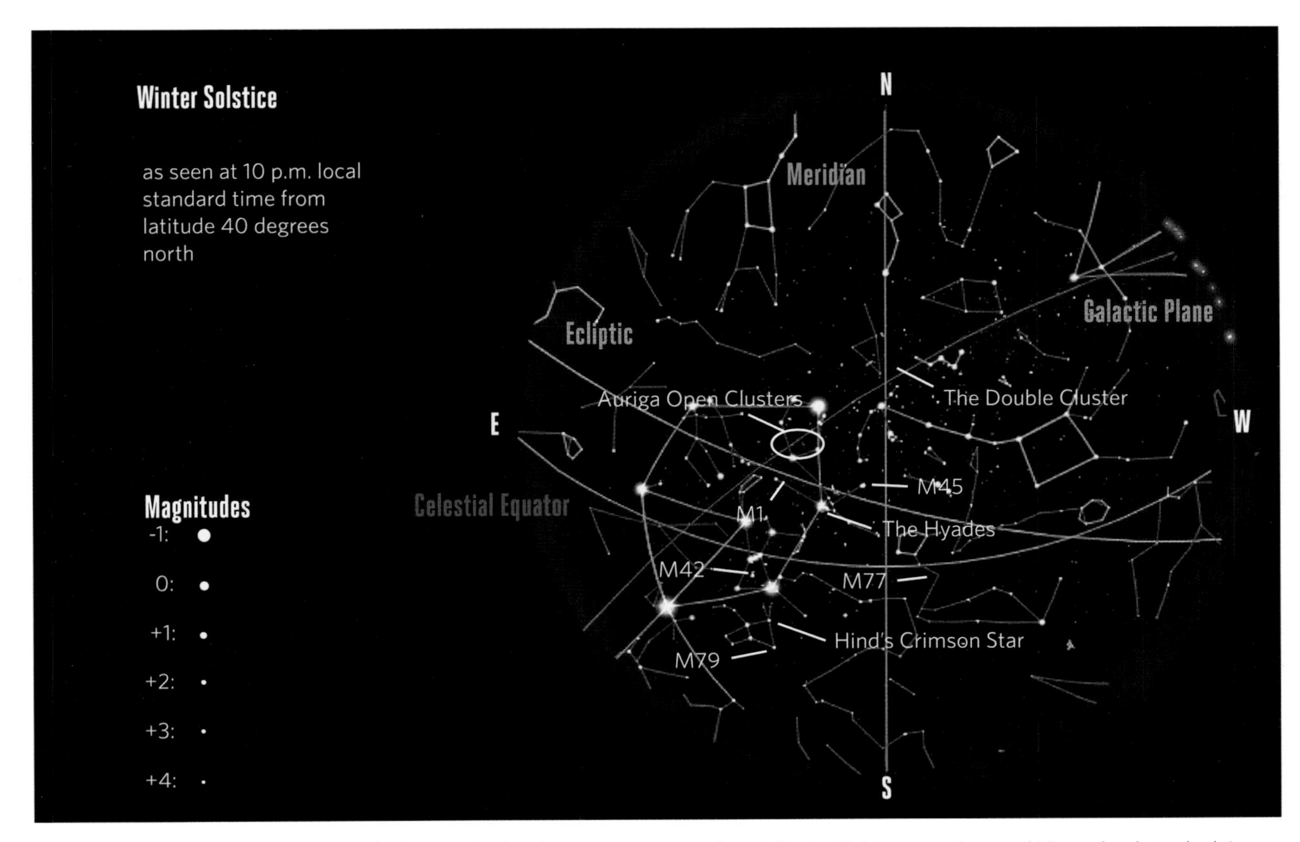

The location of the best deep sky objects (red) rising for the winter season, as seen from latitude 40 degrees north around 10 p.m. local standard time.

We can only hope that Betelgeuse pops in or around December and not in the northern hemisphere summer, with our pesky old Sun nearby. Thankfully, Betelgeuse isn't within the 50 light-year kill zone of the Earth to harm humanity (no star that has the age and mass to go supernova currently is), but will instead simply put on a great show.

At nearly equal magnitudes, here's how you can tell the twin stars of Gemini apart. Remember that "Castor with a 'C' is on the side of Capella" in the constellation Auriga, and "Pollux with a 'P' is on the side of Procyon" in the constellation Canis Minor.

The Long Nights Moon rides high on winter nights through Aries, Taurus, and Gemini near the winter solstice, occupying the space along the zodiac where the Sun will cross six months later in the summertime.

BEST OBJECTS

The Hyades: Quick: what's the closest true open cluster of stars to the Earth? That would be the V-shaped Hyades, making up the head of Taurus the Bull and located only 153 light-years away. Spilling over out of a telescopic field of view, the Hyades provide a fine sweeping view for binoculars. In 1908, astronomer Lewis Boss first noted the common proper motion of the Hyades cluster of stars toward a convergence point in Aries, moving against the nearby Ursa Major Moving Group pseudo-cluster. Hey, crosstown stellar traffic in our neck of the galaxy is rough. Ironically, bright +0.9 magnitude Aldebaran *isn't* a member of the Hyades, but instead lies 88 light-years closer in front of the cluster, making up the solitary glowing eye of the Bull. The Moon can visit the Hyades on occasion, occulting its stellar members along with Aldebaran.

Coordinates: R.A. 4H 27′ Dec. 15° 52′

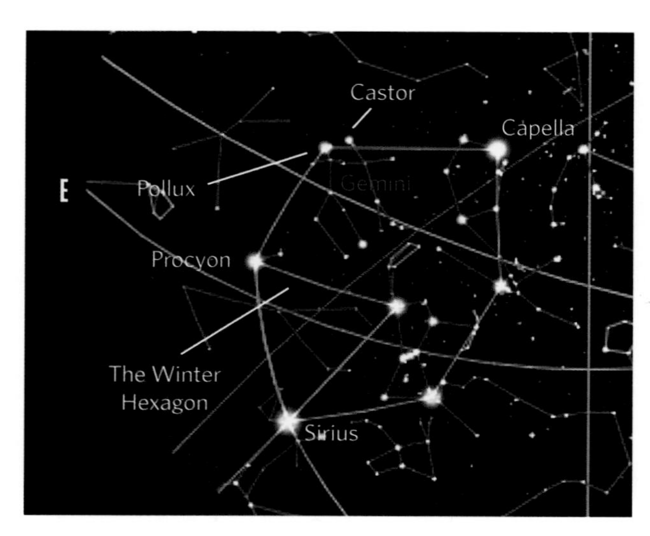

How to tell the Twin Stars of Gemini apart: Castor with a "C" is on the same side of the astronomical constellation Gemini the Twins as the bright star Capella (also with a "C") in the constellation Auriga, while Pollux with a "P" is also on the same side as Procyon in the constellation Canis Minor.

The Double Cluster: Another one that got away from Messier. The Double Cluster is a magnificent target made up of the open clusters NGCs 869 and 884. Both are physically interacting with each other, and lie 7,500 light-years away. Barely a naked eye object usually considered as between Cassiopeia and Perseus, the Double Cluster technically sits just across the border in Perseus. Spanning the breadth of two Full Moons, the Double Cluster is a surefire favorite, straddling the zenith on winter evenings.

Coordinates (Centered on NGC 869): R.A. 2H 19′ Dec. +57° 09′

M77: Though winter lacks the menagerie of galaxies visible in the spring and fall, one notable exception is M77 in the constellation Cetus the Whale. Shining at magnitude +9.6, M77 is just a degree east of Delta Ceti and easy to find. A barred spiral, M77 is also the brightest example of a Seyfert galaxy in range of a backyard telescope.

Coordinates: R.A. 2H 43′ Dec. -00° 48′

Auriga open clusters: The oblong pentagon shape of the constellation Auriga the Charioteer is brimming with a trio of fine open clusters, including (with magnitudes) M37 (+6.2), M36 (+6.3), and M38 (+7.4). Just sweep southwest of Theta Aurigae toward Taurus to pick this trio up. Can you see the contrast between tight knotty (M36) and diffuse (M38)? This set refutes the idea among deep sky observers that open clusters are all the same or boring.

Coordinates (Centered on M36): R.A. 5H 36′ Dec. +34° 08′

M42: The king of deep sky objects and an easy find, just below the celestial equator and the trio of stars in the Belt of Orion the Hunter. M42, or the Orion Nebula, makes up the fuzzy pommel of Orion's Sword, and its indistinct nature was noted since pre-telescopic times. The Orion Nebula is actually a massive star-forming region 1,300 light-years away. Look at the Trapezium—the cluster of stars at the heart of M42—and you're peering at young stars that are each much more massive than our Sun and just starting to shine and push the dusty envelope around them back, stellar ingénues making their galactic debut. This sort of cosmic drama must continue to play out billions of times over the history of the Universe.

Coordinates: R.A. 5H 35′ Dec. -05° 23.5′

M79: Lepus the Hare runs underfoot of Orion, while his faithful hunting dogs Canis Major and Minor give the hare chase on cold winter nights. Within Lepus lies a difficult but rewarding globular cluster to track down, +8.6 magnitude M79. Double the segment from Alpha to Beta Leporis south to see this challenging object.

Coordinates: R.A. 5H 24′ Dec. -24° 31.5′

M1: Known as the Crab Nebula for its pincer-shaped tendrils, this reflection nebula has a fascinating tale to tell. Eleventh-century Chinese astronomers noted a new star appearing in this patch of the sky in 1054. Today, we know that a **supernova** erupted from this region of the sky about a millennia ago. In 1968, astronomers also discovered the first **pulsar** in the Crab Nebula. M1 actually shows noticeable differences in the shape and size of the expanding nebula in images taken just a few decades apart. The Crab sits 1 degree from +3rd magnitude Zeta Tauri in a low-power field of view.

Coordinates: R.A. 5H 34.5′ Dec. +22° 01′

M45: The Seven Sisters of the Pleiades is an open cluster almost three times as distant as the Hyades. Known as Subaru in Japanese,[8] some folks erroneously think of M45 as the Little Dipper. The planet Venus can also visit M45 and perform a dramatic series of transits in front of the cluster, resuming again in April 2020.

Coordinates: R.A. 3H 47.5′ Dec. +24° 07′

Astro Challenge: Hind's Crimson Star: Lepus the Hare also houses one of our favorite star party secret weapons for winter to spring viewing. Aim at +3rd magnitude Mu Leporis, sweep 3.5 degrees to the northwest, and you're looking at a carmine-colored gem, Hind's Crimson Star. An 8th magnitude ruby-red carbon star, Hind's (also known as the variable R Leporis) is among the most colorful red stars in the sky, sure to illicit a chorus of "oohs" and "aahs."

Coordinates: R.A. 4H 59.5′ Dec. -14° 48′

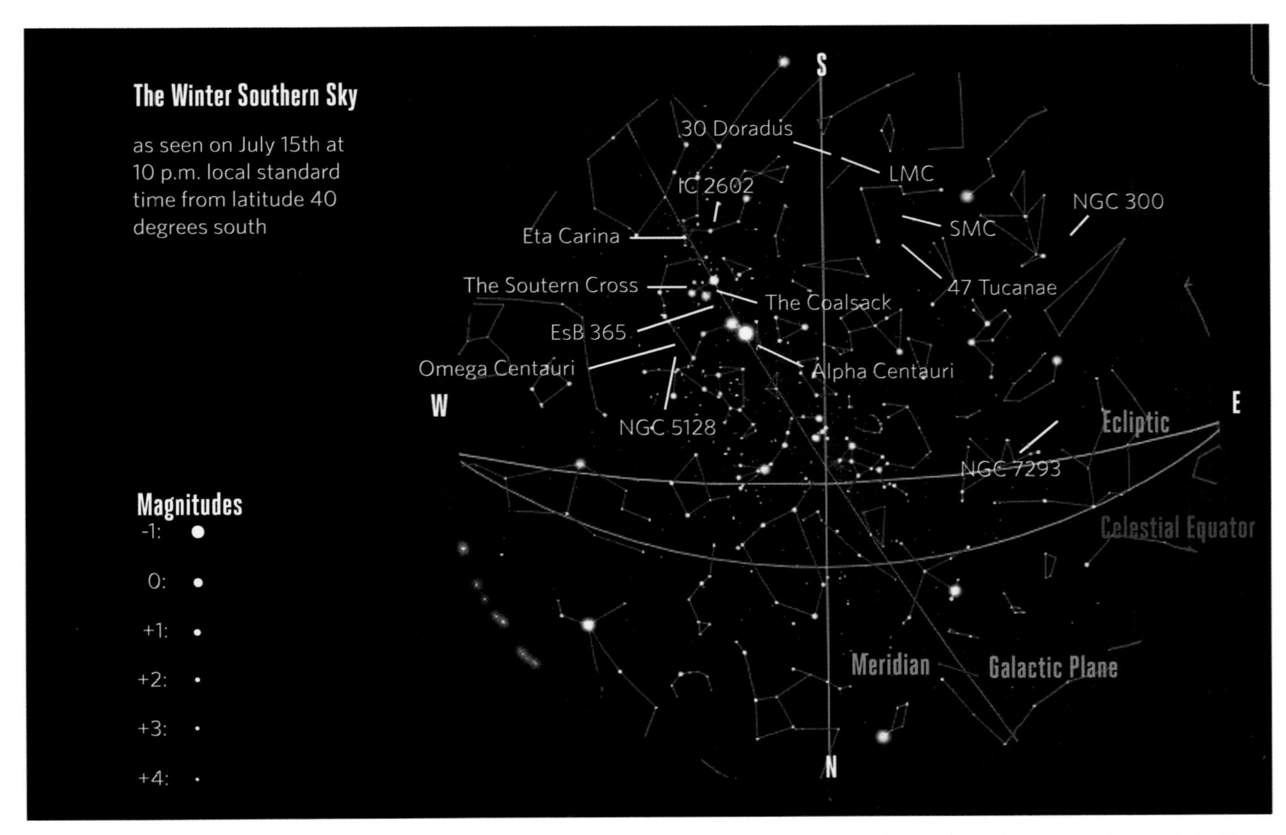

The southern hemisphere sky, as seen from latitude 40 degrees south on July 15 at 10 p.m. local standard time, along with the best sky objects for the season.

A NORTHERNER'S GUIDE TO THE SOUTHERN SKY

You finally did it. After hearing about the splendors of the southern hemisphere sky all of your life, you've finally made the plunge south of the equator. Perhaps astronomy is the sole purpose of your southern hemisphere expedition, or (more likely) you've finally managed to swing the trip as part of work, or a larger vacation or trek.

So, what's there to see? Some of the best deep sky objects lie south of the equator. I have made the journey about half a dozen times now on three different continents, and we always try to at least pack a set of binoculars for the trip. Heck, it's just strange to see familiar constellations like Leo and Orion the Hunter far to the north and appearing upside down, and to watch as familiar objects such as the Sun and the Moon rise in the east and transit to the north instead of the south.

And southern residents experience the same sort of disorientation traveling north for the first time. For example, I've heard of first-time travelers up north failing to recognize the Big Dipper, because it was simply too large!

More than likely, you have limited time to seek out the finest sights in the southern hemisphere sky. We recommend reaching out to a local astronomy club, as they will not only know their way around the unfamiliar southern sky, but they will also have access to the gear you couldn't stow in an overhead carry-on.

BEST OBJECTS

Here's our list of the best southern hemisphere objects, and when and where to see them.[9]

The Large and Small Magellanic Clouds (LMC and SMC): Ferdinand Magellan's crew made note of these fixed luminous patches deep in the southern sky during his sixteenth-century voyage around the world. About a century ago, measurements to stars in the LMC and SMC revealed that the pair are actually quite distant and are, in fact, large satellite galaxies in their own right, gravitationally bound to the Milky Way. The LMC and SMC are 163,000 and 200,000 light-years away, respectively, and both are readily visible under moderately dark skies.

Best months: November to January. Constellation: (LMC) Dorado/Mensa (SMC) Tucana. R.A.: 5H 23.5′ Dec.: -69° 45.5′ (LMC) R.A.: 00H 53′ Dec: -72° 50′ (SMC)

Southern Cross: The swaggering four bright stars of the constellation Crux or the Southern Cross are to the southern hemisphere what the iconic pattern of the Big Dipper asterism is to the northern sky. The pattern of the Southern Cross even adorns the flag of several southern nations, including New Zealand and Papua New Guinea. Though the southern sky boasts no bright relative of Polaris near its southern pole, the Southern Cross does make a handy reference point, as it points roughly near the southern rotational pole.

Best month: May. Constellation: Crux, the center of the cross. R.A.: 12H 29.5' Dec.: -59° 05'

Omega Centauri: Located 15,800 light-years away in the constellation Centaurus, Omega Centauri is the largest-known globular cluster belonging to the Milky Way Galaxy. Containing an estimated 10 million stars, Omega Centauri is a naked eye object shining at +4th magnitude and spans an area larger than the Full Moon. Astronomers theorize that Omega Centauri may be the remnants of a dwarf galaxy that our Milky Way Galaxy tore apart and largely absorbed long ago.

Best months: May/June. Constellation: Centaurus. R.A.: 13H 27′ Dec.: -47° 29′

47 Tucanae: Located very near the Small Magellanic Cloud, don't miss one of the sky's finest globular clusters, 47 Tucanae (NGC 104). Located about 13,000 light-years away, 47 Tucanae is very similar in size and shape to its only deep sky rival, Omega Centauri. Though it's near the SMC in the sky, 47 Tucanae is actually fifteen times closer to Earth. Move it up closer still to about 100 light-years away, and it would dominate the sky, with thousands of bright stars.

Best month: November. Constellation: Tucana. R.A.: 00H 24′ Dec.: -72° 05′

The Coalsack: One of the greatest sights in the southern sky represents what you *can't* see. The Coalsack Nebula sits like a black hole punched in the sky near Alpha Crux, spilling over into the southern corner of the constellation Centaurus. The Coalsack Nebula is a knotted mass of gas and dust, hiding the starry plane of the Milky Way beyond. The Coalsack is 7 by 5 degrees wide and 600 light-years away.

Best month: May. Constellation: Crux/Centaurus. R.A.: 12H 50′ Dec.: -62° 30′

Alpha Centauri: Don't miss the star system that holds the distinction as the closest to our own Solar System. Alpha Centauri is 4.4 light-years away in our current epoch. The +1 magnitude pair is an easy split at 4 arcseconds wide. Also, keep an eye out for +11th magnitude Proxima Centauri 2.2 degrees away, technically the closest star to our Solar System at 4.2 light-years from Earth.

Best month: June. Constellation: Centaurus. R.A.: 14H 39.5′ Dec.: -60° 50′

Canopus: Shining at magnitude -0.7 and 310 light-years away, Canopus, in the rambling southern constellation Carina the Keel, is the second brightest star in the sky. Draw a line straight down from Sirius in right ascension, and you run smack into Canopus. At eight times the mass of our Sun, Canopus is a monster; move it up to 10 parsecs away, and you could easily pick it out in the daytime sky at a dazzling -6th magnitude.

Best months: January/February. Constellation: Carina. R.A.: 6H 24′ Dec.: -52° 42′

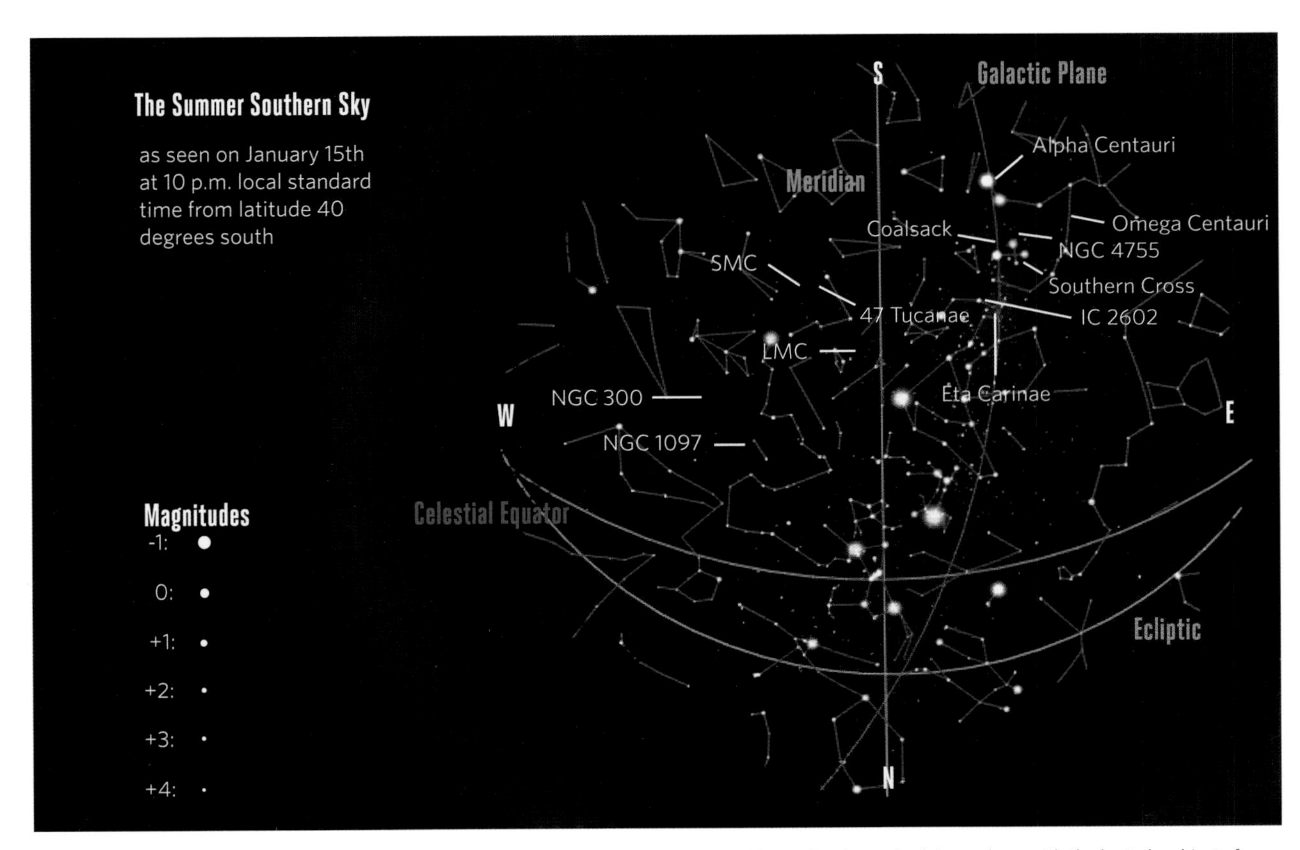

The southern hemisphere sky, as seen from latitude 40 degrees south on January 15 at 10 p.m. local standard time, along with the best sky objects for the season.

Eta Carinae: In 1843, the massive star at the heart of Eta Carinae burst forth, briefly becoming the second-brightest star in the sky. We now know that Eta Carinae hosts some of the most massive stars discovered, a pair with a combined upper mass limit of 280 Suns and 7,500 light-years away. Eta Carinae will one day go supernova, and that day could be tonight or 10,000 years from now. Eta Carinae is embedded in a cocoon of material it is currently shedding into the surrounding space, known as the Homunculus Nebula. Eta Carinae is at the core of the extended Carina Nebula, a grand emission nebula.

Best month: April. Constellation: Carina. R.A.: 10H 44′ Dec.: -59° 53′

30 Doradus: Better known as the Tarantula Nebula, 30 Doradus is one of the most prominent features in the Large Magellanic Cloud. This is also one of the most massive star forming regions, dwarfing the famous Orion Nebula. Shining at magnitude +8, the Tarantula Nebula also spans a patch of sky larger than a Full Moon. At 300 light-years across, 30 Doradus is simply massive, a fine example of an active star-forming region in its prime.

Best month: January. Constellation: Dorado. R.A.: 5H 39′ Dec.: -69° 06′

NGC 4755: For a tiny constellation, Crux contains a wealth of deep sky objects. Don't miss the finest open cluster in the southern hemisphere sky: NGC 4755. John Herschel dubbed this gem of a cluster the Jewel Box, and it's easy to see why. Located 1 degree southeast of the +1st magnitude star Beta Crucis, colorful NGC 4755 is one of the youngest open clusters known, at an estimated age of 14 million years.

Best month: May. Constellation: Crux. R.A.: 12H 54′ Dec.: -60° 22′

NGC 7293: Most planetary nebulae are little more than dots at the eyepiece. Not so with the NGC 7293, also known as the Helix Nebula. Located 700 light-years away, the Helix Nebula is the closest planetary nebula to our Solar System, and a southern analogue to the Ring and Cat's Eye nebulae up north. Shining at magnitude +7.5, the ghostly NGC 7293 spans 25 arcminutes—nearly the span of a Full Moon—and can really surprise even a seasoned observer looking for it under dark skies due to its extent . . . hey, we're just not used to seeing planetary nebulae of this proportion up north!

Best month: October. Constellation: Aquarius. R.A.: 22H 30′ Dec.: -20° 50′

NGC 5128: There are a few galaxies worth checking out south of the equator. One fine example is massive NGC 5128, often simply referred to as Centaurus A. Shining at magnitude +6.8 and spanning 25 arcminutes across, you might just be able to spy the twin lobes emanating from NGC 5128, evidence of the jets spewing forth from its core at relativistic speeds. NGC 5128 is one of the closest radio galaxies to the Milky Way at an estimated 13,000,000 light-years away, and the closest active galactic nucleus available for astronomers to study.

Best month: May. Constellation: Centaurus. R.A.: 13H 25.5′ Dec.: -43° 01′

NGC 300: One of the brightest members of the Sculptor Group of galaxies, NGC 300 sits far off the galactic plane. With an apparent magnitude of +9 covering 20 arcminutes, the face of NGC 300 is inclined nearly 45 degrees along our line of sight. NGC 300 sits 8 degrees south of the +4.3 magnitude star Alpha Sculptoris, which in turn lies very near to the southern galactic pole.

Best month: November. Constellation: Sculptor. R.A.: 00H 55′ Dec.: -37° 41′

The Fornax galaxy cluster: The southern analogues of the Virgo and Coma Berenices galaxy clusters, the constellations Sculptor and Fornax the Furnace both host galaxy clusters of their own. The highlights of the Fornax cluster of galaxies includes +9.5 magnitude NGC 1097, just 3 degrees north of Beta Fornacis.

Best month: November. Constellation: Fornax. (Position for NGC 1097): R.A.: 2H 46′ Dec.: -30° 16.5′

IC 2602: While you're sweeping through Carina, don't miss the "southern Pleiades," the Theta Carinae Cluster or IC 2602. An extended open cluster of stars, the heart of the cluster is the +2.7 magnitude star Theta Carina. The cluster itself is located about 550 light-years away, slightly farther away than the Pleiades.

Best month: April. Constellation: Carina. R.A.: 10H 43′ Dec.: -64° 24′

One more: EsB 365: It's worth cranking up the magnification on +1.2 magnitude Mimosa (Beta Crucis) in the Southern Cross to reveal a true gem: EsB 365, a +9.5 magnitude red carbon star, just 45 arcseconds away.

Best month: May. Constellation: Crux. R.A.: 12H 48′ Dec.: -59° 41′

TWENTY BRIGHTEST STARS				
Star Name	Magnitude	Constellation	R.A.	Declination[10]
Sirius	-1.5	Canis Major	6H 45′	-16° 43′
Canopus	-0.7	Carina	6H 24′	-52° 42′
Rigil Kentaurus	-0.3	Centaurus	14H 40′	-60° 50′
Arcturus	-0.04	Boötes	14H 16′	+19° 11′
Vega	0.03	Lyra	18H 37′	+38° 47′
Capella	0.08	Auriga	5H 16′	+46° 00′
Rigel	0.1	Orion	5H 15′	-8° 12′
Procyon	0.4	Canis Minor	7H 39′	+5° 13′
Achernar	0.46	Eridanus	1H 38′	-57° 14′
Betelgeuse	0.5	Orion	5H 55′	+7° 24′
Hadar	0.6	Centaurus	14H 4′	-60° 22′
Acrux	0.76	Crux	12H 27′	-63° 06′
Altair	0.77	Aquila	19H 51′	+8° 52′
Aldebaran	0.9	Taurus	4H 36′	+16° 31′
Antares	0.95	Scorpius	16H 29′	-26° 26′
Spica	0.98	Virgo	13H 25′	-11° 10′
Pollux	1.1	Gemini	7H 45′	+28° 02′
Fomalhaut	1.2	Piscis Austrinus	22H 58′	-29° 37′
Becrux	1.3	Crux	12H 47′	-59° 41′
Deneb	1.3	Cygnus	20H 41′	+45° 17′

ACTIVITY: NABBING SOUTHERN SKY OBJECTS . . . FROM UP NORTH

It's a fact about the sky we often forget. Yes, no matter where you are on the globe of the Earth, you only see practically one-half of the sky at a time.[11] However, unless you live at one of the rotational poles, you actually see *more* than half of the sky over the span of a single night . . . and this percentage gets higher as you approach the equator, where the entirety of the sky wheels overhead in the course of 24 hours.

Did you know that from latitude 30 degrees north, you can still see over three-quarters of the celestial sphere? Now, my challenge to you is to take the list of southern sky objects previous and attempt to spot several of them from up north. Start with the Helix Nebula in zodiacal constellation of Aquarius, one of the classic southern objects that just eluded Messier's catalog. Omega Centauri is also a classic example of a deep southern object visible up north, as we frequently show it off on April and May evenings as it pops up above the southern horizon from here in central Florida at latitude 30 degrees north. Near a southern declination of -47 degrees, we've heard of reliable observations of Omega Centauri as far north as 40 degrees latitude, the same latitude as the city of Philadelphia. You can also nab Gamma Crucis—the top star in the Southern Cross—right around the same time of the year.

Can you see Canopus deep down south during the northern hemisphere winter, using Sirius as a guide? How about +1.2 magnitude Fomalhaut low to the south in the fall? Can you trace out the exotic constellations Microscopium, Grus, and faint lines of Fornax? If you live south of the 40, we urge you to give it a try. You might actually find, on that (hopefully not far off) day when you finally make that once-in-a-lifetime trip south of the equator, these old friends will wait to greet you as a gateway to the glories of the southern sky.

An amazing view of the fine southern hemisphere globular cluster Omega Centauri, located 15,800 light-years away. Taken through an ED102 telescope at a focal length of 714 mm using 30-second sub-exposures with a modified Pentax K-5 DSLR camera.

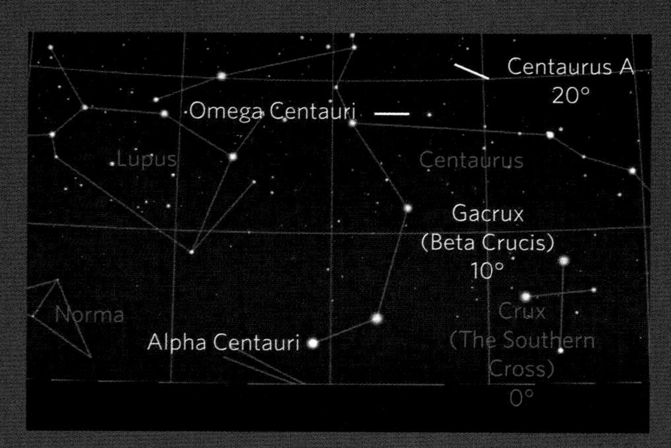

You can actually see farther into the southern hemisphere sky from the contiguous U.S. than most people realize. Here's the view from Key West, Florida, (latitude: 24 degrees, 33' north) around midnight, April 30th looking south. As you can see, famed southern targets such as the Southern Cross, Alpha Centauri, Centaurus A, and Omega Centauri just peek up over the southern horizon around this time of year. Of course, you still have to journey southward to see these objects in their true glory, high overhead.

ELEVEN BEST NAKED EYE VARIABLE STARS					
Name	**Constellation**	**Period**	**Magnitudes**	**R.A.**	**Declination**
R Aquilae	Aquila	284 days	5.5-12.0	19H 6.5′	+8° 14′
R Coronae Borealis	Corona Borealis	Varies	5.7-14.8	14H 48.5′	+28° 9.5′
Chi Cygnus	Cygnus	407 days	3.4-14.2	15H 50.5′	+32° 55′
R Leonis	Leo	310 days	5.8-10.0	9H 47.5′	+11° 26′
Beta Lyrae	Lyra	13 days	3.3-4.3	18H 50′	+33° 22′
R Scuti	Scutum	140 days	4.5-9.0	18H 47.5′	-5° 42.5′
Delta Cephei	Cepheus	5.5 days	3.4-4.3	22H 29′	+58° 25′
Mira	Cetus	332 days	2.0-10.0	2H 20′	-2° 58.5′
Algol	Perseus	3 days	2.1-3.4	3H 8′	+40° 58′
R Triangulum	Triangulum	267 days	5.4-12.6	2H 37′	+34° 16′
U Orionis	Orion	368 days	4.8-13.0	5H 56′	+20° 10.5′

TWELVE CLASSIC DOUBLE STARS FOR STAR PARTIES

Perhaps it's partly cloudy, with clear gaps seeming to pass swiftly overhead as if the sky is on a conveyor belt. Or maybe you've been asked to do a public star party from a bright school parking lot, or from a city sidewalk smack under a bright street lamp.

I've been there, trust me. While this isn't where I would *choose* to observe from on my own, I'm often asked to tease the Universe out from less-than-optimal, light-polluted skies. Nebulae and galaxies are out of the question in such a setting. Perhaps the Moon or a few bright planets are visible for show . . . or not. What do you do then? Here are some of the best bright double stars to show folks under bright urban skies.

(Note: this list is northern hemisphere-centric!)

Alcor and Mizar: A classic wide double in the bend of the handle of the Big Dipper asterism, keen-eyed observers might just be able to split the wide pair. At the eyepiece, Mizar itself is a nice double, with a split of 14 arcseconds.

Magnitude: +2.3. Separation: 12′. R.A.: 13H 23′ 56" Dec.: +54° 55′ 31″

Albireo: A classic star party double in Cygnus, known for its gold versus blue contrast. What colors do YOU see? Located 430 light-years away, there's still a controversy as to whether the pair is a true double, or merely an **optical double**, with two bright stars along nearly the same line of sight.

Magnitude: +5.8. Separation: 35″. R.A.: 19H 30′43″ Dec.: +27° 57′ 35″

Castor: What's the maximum number of stars that can exist in one solar system? With six suns, Castor in the constellation Gemini is definitely a contender. The main pair has a close split, easily resolvable at low power. A +10th magnitude star related to the system is located 73 arcseconds away. Now get this: we know from spectroscopic analysis and eclipsing variability of the fainter star that all *three stars* in the system are, in fact, doubles, making for an amazing sextuplet system.

Magnitude: +1.9. Separation: 6″. R.A.: 7H 34′ 36″ Dec.: +31° 53′ 19″

Porrima: Located in the constellation Virgo, this tight challenging binary is getting easier as it nears its widest separation of 6 arcseconds in 2070.

Magnitude: +2.7. Separation: 3″. R.A.: 12H 41′ 40″ Dec.: -01° 26′ 58″

Iota Cancri: A nice, wide double in the northern tip of the constellation Cancer, often overlooked.

Magnitude: +6.6. Separation: 30″. R.A.: 8H 46′ 42″ Dec.: +28° 45′ 37″

Epsilon Lyrae: Another classic in the constellation Lyra, very near the bright star Vega. Known as the "Double-double," Epsilon Lyrae is an easy split even at low power or with binoculars. Now, crank up the magnification and look closer. Each component is resolvable as a double, 2.3 and 2.6 arcseconds apart.

Magnitude: +4.7. Separation: 208″. R.A.: 18H 44′ 20″ Dec.: +39° 40′ 12″

40 Eridani: This triple system is a true treat. Focus in and crank up the magnification on the fainter +10th magnitude pair, which are actually a white dwarf and a red dwarf, paired together.

Magnitude: +4.4. Separation: 83″. R.A.: 4H 15′ 16″ Dec.: -7° 39′ 10″

Epsilon Boötis: A fine, close split pair.

Magnitude: +2.8. Separation: 2.8″. R.A.: 14H 44′ 59″ Dec.: +27° 4′ 27″

95 Herculis: Some doubles can play tricks on you. One such example is this pair in the constellation Hercules, which—depending on the observer—can appear green-hued, to yellow, to red. What colors do you see?

Magnitude: +5. Separation: 6.3″. R.A.: 18H 1′ 30″ Dec.: +21° 36′

Algieba: A nice, easy equal magnitude split in the curve of the Sickle asterism in Leo.

Magnitude: +2.1. Separation: 4″. R.A.: 10H 19′ 58″ Dec.: +19° 50′ 29″

Herschel 3945: Located in the constellation Canis Major, this multi-hued pair is sometimes referred to as "The Winter Albireo," and rivals its summertime namesake.

Magnitude: +5. Separation: 27″. R.A.: 7H 17′ 36″ Dec.: -23° 18′ 55″

Polaris: We sometimes forget that the famous pole star of Polaris is, in fact, a fine double. It's also the closest example of a Cepheid variable, a crucial yardstick for measuring cosmic distances.

Magnitude: +2. Separation: 18″. R.A.: 2H 30′ 42″ Dec.: +89° 15′ 38″

Omicron Cygni: A complex grouping of stars in Cygnus the Swan . . . will the real Omicron please stand up?

Magnitude: +4.8. Separation: 2′. R.A.: 20H 13′ 18″ Dec.: +46° 48′ 56″

Gamma Delphini: A fine pair, in the corner of the diamond-shaped Job's Coffin asterism in the constellation Delphinus the Dolphin. Watch for the elusive Ghost Double pair of stars Struve 2725 nearby.

Magnitude: +4.5. Separation: 9″. R.A.: 20H 46′ 39″ Dec.: +16° 7′ 38″

Gamma Arietis: A nice easy-to-find double in the tiny constellation Aries the Ram.

Magnitude: +3.9. Separation: 7.6″. R.A.: 1H 53′ 32″ Dec.: +19° 17′ 38″

REFERENCES

Messier marathon: http://www.messier.seds.org/xtra/marathon/mm-dates.html

Dark nebulae: http://www.skyandtelescope.com/observing/celestial-objects-to-watch/seeking-summers-dark-nebulae/

Messier catalog: http://www.messier.seds.org/

Interactive NGC catalog: http://spider.seds.org/ngc/ngc.html

The Best Variables in the Sky: http://www.astronomy.com/observing/get-to-know-the-night-sky/2006/12/fun-with-double-and-variable-stars

The Orion Nebula complex (Messier 42). This stellar nursery located about 1,000 light-years away is the home to massive young stars that are just starting to shine. The Running Man Nebula (M43) can be seen to the left.

NEAR-SKY WONDERS:

SATELLITES, AURORAS, AND MORE

Tracking curious changes in our atmosphere and beyond.

"Nearer the Gods no mortal may approach."

—Edmond Halley on Newton

ACTIVITIES

- What satellite is that? (page 116)
- Imaging the International Space Station (page 117)
- Aurora observing: Will there be auroras tonight? (page 121)

THROUGH SHEER HUBRIS,
we've altered the night sky itself.

Modern civilization has finally done it. Through sheer hubris, we've altered the night sky itself. In previous chapters, we examined the sky out to its very most distant cosmic outskirts. Now, let's zoom back down to what's skimming the atmosphere, just overhead. Watch the sky in the twilight hours right after sunset or early in the dawn just before sunrise, and you might just spy a starlike point of light sliding silently by. This is an **artificial satellite**, a testament to the modern Space Age. The era of artificial satellites began on October 4, 1957,[1] with the launch of Sputnik 1.[2] Back then, the world met the news of the first artificial moon in orbit with both wonder and fear. This was the height of the Cold War, and many feared the runaway nuclear arms race would soon head into space. In the U.S., there was also realization of a nagging technology gap, highlighting an urgent need to spearhead science education to keep the nation technologically competitive.

The U.S. entered the Space Race on January 31, 1958, with the launch of Explorer 1. Today, the United States Joint Space Operations Center (US JSpOC) lists over 43,000 orbital objects launched over the last 60 years. The US Space Surveillance Network follows an estimated 23,000 objects remaining in orbit around the Earth larger than 5 centimeters in size.[3] These date all the way back to Vanguard 1, which was launched on March 17, 1958, and was expected to remain in space for another 180 years. Objects in orbit range in size and shape from the International Space Station—which is nearly as big as an American football field at 356 feet (108.5 m) along its longest axis—to CubeSats, shards of satellites, and small pieces of debris.

Amateur satellite watchers were also on hand from the very beginning, playing a crucial role in chronicling the early Space Age. One such early effort was Project Moonwatch. Sputnik caught the U.S. off guard, before planned optical tracking stations were up and running. Project Moonwatch sprang from earlier Civil Air Patrol programs training volunteer spotters to watch for incoming Soviet bombers. Project Moonwatch enlisted and trained teams of volunteers to track and time satellites.

The International Space Station as seen from orbit from the U.S. Space Shuttle Discovery.

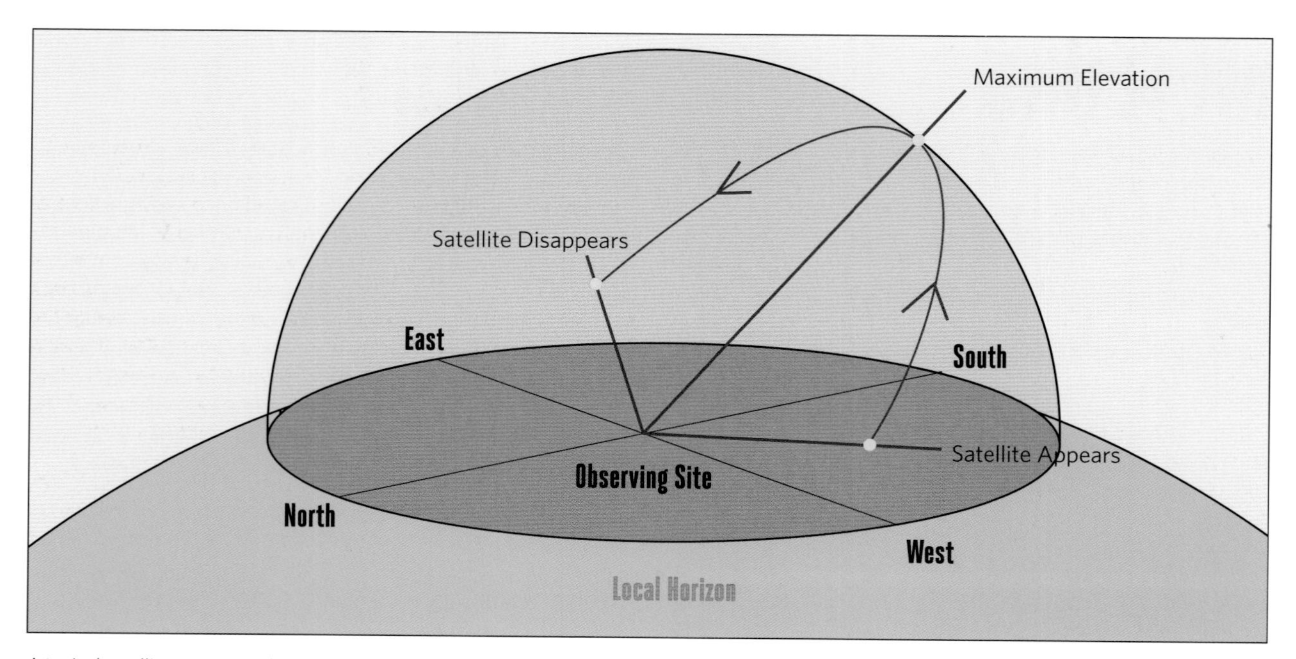

A typical satellite passes overhead, using the "sky as a bowl" illusion over your local observing site. Most satellites will appear from low in the west as they clear the thick murk of the atmosphere low to the horizon, then appear highest and brightest as they transit the local meridian a few minutes later, then fade out and disappear low to the east. Note that satellites also appear to change speed as they move across the sky, moving slow low to the horizon as they approach the observer, then faster as they pass directly overhead. Only polar satellites in retrograde (backward) orbits defy this trend, moving in highly inclined orbits from east to west.

Though Project Moonwatch formally ended in 1975,[4] a network of dedicated volunteers continues to carry on this effort worldwide. These amateur satellite spotters perform a valuable service, keeping tabs on launches and clandestine payloads, foreign and domestic. These sorts of observations can confirm or deny rogue nation claims of successful launches and track changes or technology tests in space.

Satellite, aircraft, or other? As with meteor showers, one of the best aspects of chasing satellites is its relative simplicity. You simply have to watch the right area of the sky at the right time. Most of the naked eye satellites you're seeing at dawn or dusk are in **Low Earth orbit** (LEO), traveling swiftly around the planet once every 90 minutes. LEO is surprisingly close: only around 200 miles (322 km) away—an afternoon's drive, if you could somehow drive your car straight up.

But how can you be sure that your suspect moving star is really a satellite? Here are some quick and dirty tips.

Is it blinking? Satellites shine because they're still catching the last rays of the Sun, high overhead. Incidentally, this is yet more visual evidence that the Earth is indeed round; while it's getting dark from your observing site and the Sun has long since set, sunlight is still streaming past the curve of the planet through the realm of the satellites, high overhead. True, some satellites can vary in brightness and flare as their reflective solar panels catch the Sun just right, but a slow flaring defers visually from the quick, rhythmic blinking of aircraft anti-collision lights. A good rule to remember is if a slow-moving light blinks, it's an aircraft, if it stays pretty much steady and slowly fades out when it hits the Earth's shadow, it's a satellite, or more likely a rocket booster from a previous mission, still in orbit.

METEORS

The glowing trail of a meteor is iconic and is usually pretty tough to confuse with anything else. Very occasionally, you might spy the flash of a pinpoint meteor, moving directly at you along your line of sight. Early morning meteors also move faster than stately meteors hitting the Earth before midnight. This is because you are on the forward-facing side of the Earth past midnight, with incoming meteors hitting the Earth head on. Any meteors hitting the Earth from behind in the evening have to race to catch up.

Bright meteors can also leave lingering glowing trails for several minutes after passage. Very occasionally, one might witness a slow **meteor train** (also known as a **meteor procession**), as a pea-size or larger space rock breaks up. When is a meteor train really a satellite? When it's actually space junk, reentering the Earth's atmosphere. Though NASA and the Russian space agency Roscosmos directs many large reentries toward a remote southern hemisphere location referred to as **Point Nemo** at the end of their utility, most reentries are simply uncontrolled.[5] When footage of a bright meteor makes its way around the web, satellite sleuths generally pin it to a suspected reentry within a few hours. Satellites also generally have much slower reentry velocities than meteors, making for a long drawn-out spectacle lasting up to a minute or so. And just perhaps, a satellite reentry may put on a colorful blue-green show due to metals such as copper or aluminum wiring burning up, putting on one last Bunsen-burner flame test straight out of high school chemistry class through the night sky.

The location of the "Satellite Graveyard," Point Nemo, over the South Pacific.

WILDLIFE

It might seem silly . . . until you've seen it. Nocturnal animals seen under extraordinary conditions can occasionally mimic moving glowing lights in the sky. I recall once seeing a flock of migrating birds high up at deep dusk, streetlights illuminating their undersides as they flew past. It looked pretty convincing as a set of moving "stars" high overhead . . . until I took a closer look with binoculars and saw the flapping wings causing the lights to wink on and off. Bats also like to come out at dusk and dive bomb observing sites looking for insects. Flashing fireflies also sometimes turn up on astrophotography images of the night sky, the chemical luciferin illuminating their abdomens.

Insects and pollen can also flash in pearly halos near the Sun, Moon, and nearby streetlights, in a phenomenon known as *lichtflocken* (German for light flecks).[6] More pedestrian UFOs actually turn out to be weather balloons or even empty garbage bags swept aloft, catching the glint of the twilight Sun.

A good observer is always skeptical, and once you know your way around the night sky, it's old hat to run through the mental checklist of satellite versus not a satellite.

TYPES OF ORBITS

Satellites are placed into various types of orbits, depending on their mission. Most Low Earth orbit satellites follow a simple west-to-east path, mimicking the rotation of the Earth. This sort of orbit is also the easiest to achieve, as you can use the boost from the Earth's rotation to get your prospective satellite on its way.[7] Target a given azimuth, and you can put your satellite in a desired **inclination**, or the tilt of its path in degrees relative to the Earth's rotational equator. The International Space Station, for example, orbits the Earth in a high 51.6-degree inclination, allowing access from space ports worldwide. Satellite orbits also follow nearest and farthest points from the Earth, known respectively as **perigee** and **apogee**. Contrary to what you see in Hollywood, orbits are basically fixed, though they are subject to the effects of solar wind pressure and atmospheric drag. Think of all that energy the very first disposable rocket stages put into a given launch. That's lots of energy to bleed off, and Newton's third law of motion tells us it would take an equivalent amount of energy to take an object back down.

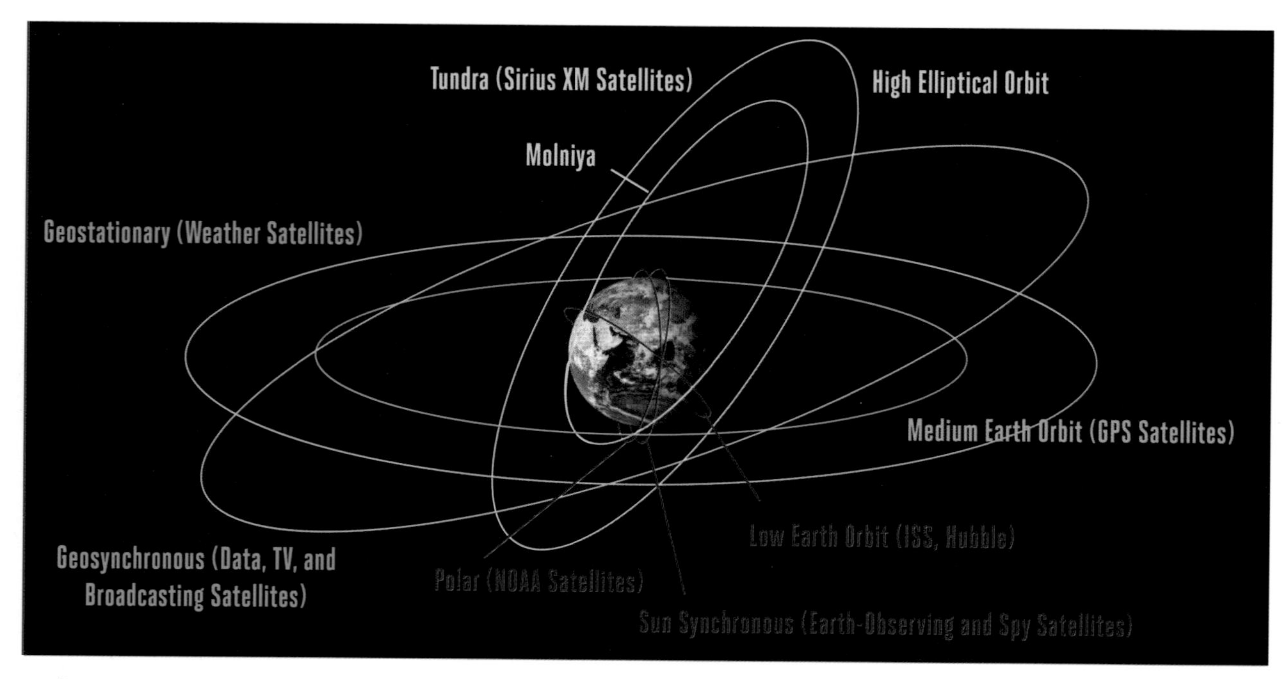

An illustration of the types of Earth orbits described in the text, with specific examples of each: Low Earth Orbit (LEO, red), Medium Earth Orbit (MEO, green), High Elliptical Orbit (HEO, yellow), and Geostationary/Geosynchronous Earth Orbit (GEO, blue). Orbits versus the disk of the Earth are to scale.

NASA and US JSpOC describe the orbital elements of a satellites using what's known as **two-line elements** or TLEs. These elements are published publicly for non-classified missions shortly after a successful launch reaches orbit with a NORAD COSPAR identifier (named for the international COuncil for science's committee on SPAce Research),[8] which designates the year and the order of launch. For example, the ISS is "1998-067A," denoting the launch of the Zarya core module as the sixty-seventh successful launch in 1998. The final letter "A" denotes the first object in orbit for a given launch, followed by "B, C," etc., for successive payloads. Keep in mind, success in rocketry means it reached orbit . . . it may, like the ill-fated Japanese Hitomi X-ray observatory, fail once it's in orbit.

Satellites are also assigned a Satellite Catalog Number once in orbit. This is simply a linear chronological number, going all the way back to 00001, the Sputnik rocket booster that started it all.

Another special type of Low Earth orbit is a **sun-synchronous**, or Earth-observing orbit. This is a retrograde or backward path, allowing the satellite to move from east-to-west against the Earth's rotation. This is handy when looking back at the surface of the planet, as the cameras will pass over the same site with the same Sun angle, making the image interpreter's job much easier. This type of orbit is handy for science satellites looking for seasonal and terrain changes . . . and spy satellites.

Farther out, satellites are placed in **Medium Earth orbits** (MEO). These are highly elliptical. Some are simply transfer orbits for boosting satellites into geostationary orbits. The Russians pioneered a type of semi-synchronous path known as a **Molniya orbit**, which allows a satellite to loiter or spend most of its time over one geographical spot near apogee and return to this point in its long elliptical orbit once every 12 hours.[9]

Finally, there's a point 22,236 miles (35,785 km) up where a satellite stays in lock step with the 24-hour rotation of the Earth below. Back in 1945, science fiction writer Arthur C. Clarke proposed placing three satellites here for global communications coverage. Today, we call this zone the **Clarke Belt**. Place a satellite here at zero-degrees inclination, and it's in a **geostationary orbit**. Do the same but tilt its orbit, and it's in a **geosynchronous orbit**. These are handy for weather and telecommunications satellites, which are generally slotted over a desired longitude for the customer.[10]

```
Space-Track Catalog Number     COSPAR ID   Date Generated (Epoch)                  Element Set Number

1 20580U 90037B   18057.60755618 +.00000469 +00000-0 +17807-4 0  9999
2 20580 028.4694 126.0723 0002880 034.9696 084.6771 15.08985984328661

"U" (Unclassified)   Inclination     Orbital Elements (eccentricity, mean anomaly, etc.)
```

A satellite Two-Line Element breakdown. "Catalog Number" goes back to "00001" (Sputnik), COSPAR ID is the last two digits of the year plus the sequential launch number for that year, then the letter object for that launch, and "epoch" is the last two digits of the year generated plus the Julian day. Incidentally, this particular TLE is for the Hubble Space Telescope, the "B" object released from the payload bay of space shuttle Discovery (1990-037A, the 37th successful launch to reach orbit worldwide in 1990) on STS-31.

When a geostationary/geosynchronous earth orbit (GEO) satellite reaches the end of its service life, it often uses the remainder of its maneuvering fuel to enter a **graveyard** or **super-synchronous** orbit a few hundred kilometers above its original operational orbit.

You can't usually see satellites in GEO with the naked eye unless they flare . . . but they do turn up on long exposure photographs. They're also occasionally visible with binoculars or a telescope, their slow nodding south-to-north and back again motion against the starry background betraying their presence. The area around the Orion Nebula (M42) is notorious for geostationary satellite sightings.

Another rare sort of super-synchronous orbit is known as a **Tundra orbit**. This also orbits the Earth on a highly elliptical path, looping in a figure eight–track over the surface of the Earth in one 24-hour sidereal day using a phenomenon known as **apogee dwell**. Early Sirius-XM satellites were placed in Tundra orbits, and are just visible to the naked eye near perigee. As with Molniya orbits, the key advantage is to provide geostationary service to higher latitudes.

The spectacular predawn launch of MUOS-4 from Cape Canaveral in September 2015. Often the most spectacular launches occur near dawn or dusk, as the rocket plume hits the sunlight streaming high over the curve of the Earth. Shot from Hudson, Florida, 100 miles (160 km) west of the launch site. The bright "star" near the launch plume is Venus.

SEEING LAUNCHES

Want to see the source of all those satellites? Good news: Seeing a launch from where you live might be easier than you think. Again, the key is to look in the right direction, at the right time. I used to live about 100 miles (160 km) west of Cape Canaveral and the Kennedy Space Center, and I could see a launch on average every month. Night launches are especially prominent. There are three orbital launch sites currently active in the U.S.: the Kennedy Space Center/Cape Canaveral launch complex on the Florida Space Coast, Wallops Island in Virginia, and Vandenberg Air Force Base in California. ISS-chasing launches going up the U.S. Eastern Seaboard often put on quite a show, with a twilight encore over the UK and Europe about 20 minutes after launch. Predawn launches are dramatic, as the engine plume often lights up the dawn sky like a luminous flower. High-altitude scuds of lingering neon-blue **noctilucent clouds** may also persist up to an hour after launch.

At Cape Canaveral in Florida, watching a launch from touristy Cocoa Beach is almost as good a vantage point as actually going out on to the Cape. Another free spot favored by locals is the top of the causeway bridge in Titusville.

EXTRA: ACTIVE SPACE LAUNCH SITES AROUND THE WORLD

Live near one of these sites worldwide? The chance to see a rocket launch could be closer than you think. This list only includes sites that currently conduct orbital space launches, or those of historical interest that did so in the past.

In this modern age, you can also follow nearly every launch worldwide, live online. NASA, the European Space Agency, JAXA, and the Indian Space Agency ISRO webcast all launches in near real-time. SpaceX now routinely broadcasts launches of its Falcon rockets, right through subsequent landing of the first stages. The Russian Space Agency Roscosmos broadcasts all launches except government/military missions live, and China occasionally broadcasts crewed and high-interest space exploration missions on its national carrier, CCTV. Smaller states such as Israel, Iran, and North Korea carry out some of the only unannounced/secret launches these days. It's amazing to watch webcasts from the historic Baikonur Cosmodrome in Kazakhstan live on the web, as these were all carried out in secret during the Cold War, just a few short decades ago.

SPACE LAUNCH SITES WORLDWIDE

Site	Country	Operator	Location	Status
Kennedy Space Center	USA	NASA	Florida	Active
Cape Canaveral Air Force Station	USA	NASA	Florida	Active
Wallops Island	USA	NASA	Virginia	Active
Vandenberg AFB	USA	U.S. Air Force	California	Active
Baikonur Cosmodrome	Russia	Roscosmos	Kazakhstan	Active
Kodiak Launch Center	USA	NASA	Alaska	Active
Kwajalein Test Range	USA	NASA/U.S. Air Force	Kwajalein Atoll	Active
Plesetsk Cosmodrome	Russia	Roscosmos	Russia	Active
Vostochny Cosmodrome	Russia	Roscosmos	Siberia	Active
Yasny Cosmodrome	Russia	Roscosmos	Siberia	Active
XiChang Launch Center	China	CNSA	China	Active

(continued)

The dramatic night launch of a SpaceX Falcon-9 rocket from Cape Canaveral Air Force Station in Florida on the night of January 7, 2018, with the classified Zuma payload. Photographer Glenn Davis notes: "This photo is a long exposure composite, consisting of fourteen 30-second frames." Location: Jetty Park, Florida, Camera: Sony a6000, Lens: Rokinon 12mm f/2.0, Aperture: f/11, Shutter Speed: 30 seconds at ISO: 100. The second streak bisecting the launch arc is an exposure composite of the SpaceX Falcon-9 stage 1 booster returning for landing.

SPACE LAUNCH SITES WORLDWIDE				
Site	**Country**	**Operator**	**Location**	**Status**
Jiuquan Launch Center	China	CNSA	China	Active
Taiyuan Launch Center	China	CNSA	China	Active
WenChang Launch Center	China	CNSA	China	Active
Satish Dhawan	India	ISRO	India	Active
Kourou Space Center	ESA	ESA	French Guiana	Active
Ocean Odyssey Launch Complex	Pacific	Sea Launch	Mid-Pacific Ocean	Active
Tanegashima Space Center	Japan	JAXA	Japan	Active
Naro Launch Center	ROK	ROK Space Agency	South Korea	Active
Sohae Launch Center	North Korea	North Korea	North Korea	Active
Pamachim Air Base	Israel	Israeli Air Force	Israel	Active
Semnan Space Port	Iran	Iranian Space Agency	Iran	Active
Hammaguira Test Center	Algeria	CNES/ France	Algeria	Inactive
Woomera Launch Center	Australia	UK Space Agency	Australia	Inactive
Broglio Space Center	Kenya	Italian Space Agency	Kenya	Inactive
Rocket Lab Launch Complex 1	New Zealand	Rocket Lab	New Zealand	Active
Spaceport America	USA	Virgin Galactic	New Mexico	Future Site
South Texas Launch Site	USA	SpaceX	Texas	Future Site
Nova Scotia Launch Facility	Canada	CSA	Nova Scotia	Future Site

Watch the sky long enough, and a satellite (or two) may photobomb the view, such as these time-lapse streaks seen on this image of the Rosette Nebula. Located just north of the celestial equator in the constellation Monoceros the Unicorn, the Rosette is frequently crossed by satellites in geosynchronous/geostationary orbit. The stops in the trails are interruptions in the CCD readouts, not tumbling motions of the satellites themselves.

TUMBLERS AND FLASHERS

As mentioned previously, one type of satellite phenomenon *does* break the "aircraft flash, satellites don't" rule. Occasionally, you'll see satellites flare brightly as they catch the glint of the distant Sun. This effect is often dramatic and rhythmic if a satellite or booster is tumbling end over end. The first Iridium constellation of satellites are notorious for this.[11] The International Space Station, the Hubble Space Telescope, and the COSMO-SkyMed series of satellites can all flare in a similar fashion.

Many interesting satellites never break naked eye visibility.

CATCHING SATELLITES WITH BINOCULARS

Many interesting satellites never break naked eye visibility. Nearly every observer has had the experience of checking out a nebula or cluster through a telescope and seeing a surreptitious starlike point of light cross the field of view, as a satellite zips by. It's getting crowded up there, indeed. My tried-and-true technique for hunting down fainter satellites is to simply note when they'll pass near a bright star, then sit back in our lawn chair with binoculars and watch at the appointed time.

Heavens-Above will show the illuminated pass of a given satellite for your location against a star map background. I also like to have WWV shortwave radio (broadcasting of AM shortwave frequencies 5, 10, and 15 MHz) for an audible and precise time hack in the background. By doing this, you can keep your eyes constantly on the sky.

I've used this method to track down the old Vanguard satellites, the crippled Hitomi X-ray observatory, North Korea's Kwangmyŏngsŏng-4 satellite, Canada's first satellite Alouette-1, and the tool bag lost by ISS astronauts in 2009.

HIGH-INTEREST SATELLITES

Track satellites long enough, and you might discover a certain favorite niche of your own. Some satspotters like chasing after clandestine spy satellites; others make a sport out of recovering recently launched payloads, or looking for older historic satellites still up there in orbit. To our knowledge, no organization gives out an equivalent Messier-type award for completing a particular class of satellites . . . and working your way through NORAD's entire catalog would be a formidable, lifelong task.

Here are ten high-interest satellites to pique your interest. Consider this list a sort of gateway drug into a highly addictive (but healthy!) pursuit of satellite-spotting.

IMAGING SATELLITES

Imaging a satellite pass is as easy as setting a DSLR digital camera on a sturdy tripod with a wide field (18–35 mm) lens, opening the aperture up as wide as it will go, and taking a time exposure shot while the satellite is in view. The satellite will look like a tiny streak through the star field. You can usually get away with a 15-second wide-angle exposure without tracking, after which, the stars will start trailing as well. Use the Manual setting on the camera, focus on a bright star, then wait until the satellite is in view. Tumblers or flashing satellites—such as the first-generation Iridiums—can add a dramatic flare to the composition. Use a manual or time-delayed shutter release, or you'll end up with a little jiggle at the very beginning of the exposure. Another way to beat this is the old hat trick method: physically cover (but don't touch) the lens, trip the shutter to start the time exposure, then uncover the lens, exposing it to the sky.

It's also possible to image a satellite through a telescope . . . and get actual detail. (See Imaging the International Space Station on page 117).

TEN HIGH-INTEREST SATELLITES TO WATCH FOR

Name	ID	Inclination	Max. magnitude	Notes
International Space Station	1998-067A	51.6°	-5.8	Crewed since 2000
Tiangong-2*	2016-057A	42.8°	-0.9	Second Chinese space station
Hubble Space Telescope	1990-037B	28.5°	-0.5	Visible from lat. -35°N to S
OTV-5	2017-052A	54.5°	+2?	USAF's classified X-37B
Iridiums	Ninety-five satellites	Var.	-8	Final reentry set for 2019?[12]
COSMO-SkyMed	Four satellites	97.8°	-3	Sun-synchronous orbit
LaCrosse-5	2005-016A	57°	2	Vanishing spy satellite
Envisat	2002-009A	98.2°	2.5	Earth-observing satellites
LightSail-2	N/A	N/A	-4?	Launching in 2018
Telkom-3	2012-044A	49.9°	-0.4	Tumbler in GEO xfer orbit

* China may also launch the core module for its Tianhe-1 space station in 2018.

TRUE ODDITIES

Looking for something more? Watch the sky long enough and you might see something truly odd. Discussed below are some of the stranger satellites out there that I'm sure are the cause of more than a few UFO sightings over the years.

LASER SATS

Some Earth-observing satellites use active Light Detection and Ranging (LIDAR) for atmospheric and elevation measurements. When these come directly over your position, they can become briefly visible as a +2 magnitude green flashing point of light. Don't worry—the laser is dispersed, and safe to look at from a 200 miles (322 km)-plus distance. One notorious laser satellite is CALIPSO. NASA's Cloud Aerosol Transport System (CATS) aboard the International Space Station should also be visible from the ground during dark shadow passes of the ISS.

NOSS PAIRS

Back in the early days of electronic maritime navigation, the U.S. launched groups of satellites as part of its Naval Ocean Surveillance System (NOSS). Some of these are still up there, silently moving in groups of two to three satellites, looking like mobile constellations in the sky. Recently, China has gotten in to the free-flying satellite game, with its Yaogan series of naval surveillance satellites.[13]

GEOSAT FLARE SEASON

Most of the time, satellites way out in geosynchronous orbit are faint and invisible to the naked eye. One exception to this rule is near either equinox, when **geostationary satellite flare/eclipse season** begins. Right around the weeks leading up to either equinox in March and September, satellites in geostationary orbits cross into the shadow of the Earth, often flaring into visibility on egress or ingress as they near the Earth's shadow and reach 100 percent illumination. Looking sunward, satellite operators also expect radio outages around this time, as radio interference from the Sun swamps satellites in GEO daily . . . another good reason to use an inclined or Tundra-style orbit.

If you monitor space launches worldwide, you might also manage to catch crewed and uncrewed missions as they arrive at and depart from the International Space Station. These will follow along the same path as the ISS, looking like a faint starlike point trailing in front of or behind the ISS. Russia's Progress, JAXA's HTV, ESA's ATV, Orbital Science's Cygnus, and SpaceX's Dragon capsule can all carry cargo to the ISS. Dragon also has the only automated downmass capability, as it can also reenter and splashdown, returning cargo and experiments to Earth. The Russian Soyuz is currently the only ride to get astronauts and cosmonauts to and from the station, though crewed Dragon flights may start in late 2018 or 2019. Unfortunately, none of these are as bright as the U.S. Space Shuttles, which ended service in 2011.

See the Resources section on page 226 for a complete list of satellite-tracking resources.

Happy satellite hunting!

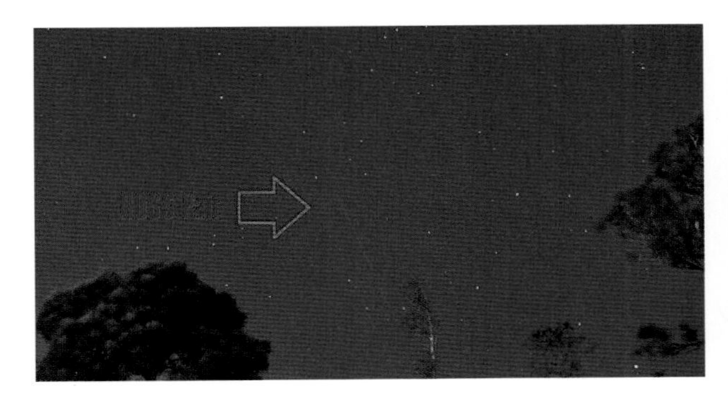

A time lapse capture of a NOSS pair of satellites.

ACTIVITY: WHAT SATELLITE IS THAT?

It's a common question I often get online and out observing. "I saw a moving light tonight, obviously a satellite . . . but *which* one?" Generally, if you can provide me with a few key pieces of information, I can sleuth it out in a few minutes . . . or at least give you a high-confidence best guess.

In this activity, we'll show you how to do this for yourself. When you see a suspect satellite, be sure to note the following elements.

Location, location, location—in latitude and longitude, down to an accuracy of at least a few arcseconds, which equals about 102 feet (31 m) at the equator. These days, a good compass app or the GPS app on your smartphone can supply this level of precision. The Heavens-Above app gives a good GPS location readout.

What was the time when you saw the satellite, down to within about 10 seconds? Again, WWV radio is a good source for an audible time hack . . . these days, more and more people are simply using their phones to tell time. If your phone or timekeeping source is automatically synced to the internet, you're all set; otherwise, I'd take the recorded time from a wristwatch and check the offset against an online time source such as Heavens-Above's "What Time is It?" link on their main page.

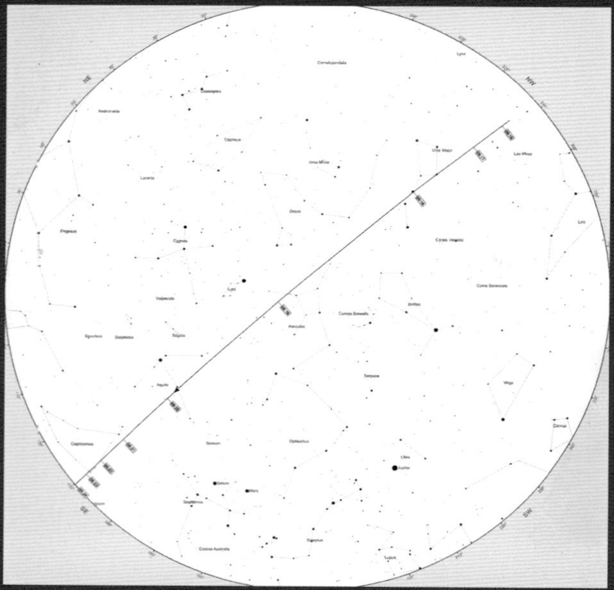

A typical satellite pass chart from Heavens-Above, showing the time and the pass of a satellite (in this case, the International Space Station) against the sky.

How bright was it? Note the brightness to the nearest full magnitude, using a nearby comparison star and a star chart.

What direction was it traveling? Here's where it's handy to know your constellations. I'd note a given pass as, for example, "Moving north to northwest through the Bowl of the Big Dipper."

Got all that? Now, I'm going to show you a nifty trick. On the main page of Heavens-Above, there's an entry for "Daily Predictions for Brighter Satellites." The user can even filter this for a given brightness cutoff. Be sure your location is set for your observing site in your profile. It will then kick out a list of satellite passes for a given date from the drop-down window, either evening or morning. From there, it's just a simple matter of matching up your suspect satellite observation with a given time and path. Click on the highlighted link marked "Time" for any given satellite to see its path laid out against the starry background.

When someone supplies me with the correct info, I can generally use this simple method to identify what satellites they saw with an accuracy of about 90 percent. Of course, once in a while, the satellite simply remains a mystery. There's lots of classified stuff up there, and sometimes the observer might've caught a fainter satellite under unusual conditions, making it appear brighter than normal.

I can generally solve the "what satellite is that?" puzzle, and now you can, too.

ACTIVITY: IMAGING THE INTERNATIONAL SPACE STATION

It's the next logical step. Nearly every time I'm showing someone a bright satellite pass at a star party, the question comes up: "Can you aim the telescope at that thing?"

The answer is yes, though it's tough to track a moving object in orbit. But you can easily image detail on the International Space Station with a little bit of practice.

Three hundred and fifty feet (107 m) from tip to tip, the International Space Station is the largest object ever constructed in orbit. At the eyepiece, the ISS spans about 51 arcseconds when it's directly overhead, larger than the angular span of Saturn, rings and all. This means you can easily see the structures of the trusses and enormous solar panels using an optical aid. Depending on how it's oriented, the ISS can appear either wafer-thin or box-shaped. Using binoculars, I can just spy the ISS as having a miniature Star Wars TIE Fighter–shape on good overhead passes.

There are sophisticated programs for satellite tracking, but here's a very low-tech but effective approach. We simply mount our video system afocally (that is, with the camera simply looking into the eyepiece, rather than mounted to the telescope *in place of* the eyepiece) through the eyepiece of the telescope, set it running as the ISS approaches, then try to keep the station as centered as possible through the Telrad reticle view. On playback, I'll usually have a few sharp images of the station as it dances briefly through the field of view.

One key to steady manual guiding is the use of a long steering arm attached to the tube of the telescope. A stick or a pool cue would work fine. I use a long pot grabber attached with bungee cords to the body of the telescope. Think of yourself as a World War II triple-A gunner or a cameraman, guiding the scope aimed at the station as it passes overhead.

Set up a few minutes prior to the pass, and be sure of the following.

- Your finder is properly aligned.
- The image is focused.
- The contrast on the camera is adjusted accordingly . . . the ISS is dazzlingly bright! You'll generally have to go through the control panel settings to adjust the brightness and contrast.
- The cover of the telescope is off and the camera is actually running (it sounds silly, I know, but it happens more than you'd think in the busy minutes leading up to an ISS pass).

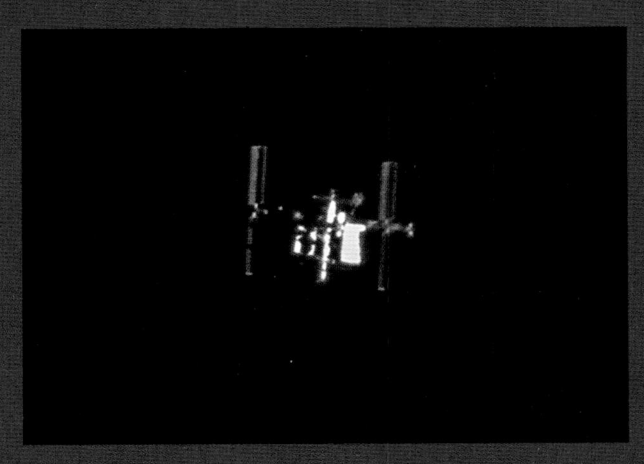

An amazing capture of the International Space Station showing structure and detail.

Venus (if it's visible, or Jupiter or a bright star if it's not) makes a great target to set up the alignment, focus, and contrast on. It's always worth practicing on an aircraft or two if possible.

You can also use the ambush method and simply stake out a bright star where you know that the ISS will pass, set the video running, and wait.

Another interesting variation of this method is to catch the ISS transiting or passing in front of the Sun or Moon. In the case of solar transits, you'll need a telescope or camera zoom lens with a proper safe solar filter (see Chapter 8: Solar Observing, page 140) and you won't see the ISS approaching the Sun until it briefly passes in front. In the case of the Moon, the ISS may or may not be visible as it approaches. At only half a degree across, transits of the ISS in front of the Sun or Moon are *quick*, generally lasting less than a second. Precise timing is essential. A great resource for this quest is CALSky, which will generate email alerts for your location when a transit is upcoming. Another excellent site is Transit Finder: http://transit-finder.com/. The great thing about using Transit Finder for imaging the International Space Station is that you don't need to track it. It lets you know where the ISS is going to be, so you just need video running of the Moon at the correct time.

Always use as high a frame rate as possible. The faster you're shooting, the more images of the ISS will turn up in post-processing.

Good luck!

OBSERVING AURORAS

On September 1, 1859, the astronomer Richard Carrington was making a daily sketch of a massive sunspot region when the area erupted in a burst of light: "Being somewhat flurried by the surprise . . . I hastily ran to call someone to witness the exhibition with me," Carrington wrote. "On returning within 60 seconds, I was mortified to find that it was already much changed and enfeebled."[14] Carrington had just witnessed a powerful **white light flare**, one of the most massive and energetic displays ever recorded on our Sun. Within days, telegraph operators began to feel the effects, as the extra energy overloaded transmission wires and set at least one telegraph office aflame. Alarmed operators soon discovered that there was enough ambient current on the line to transmit signals while disconnected from a power source, and word quickly flashed up and down the U.S. Eastern Seaboard for other operators to follow suit and disconnect their precious batteries.

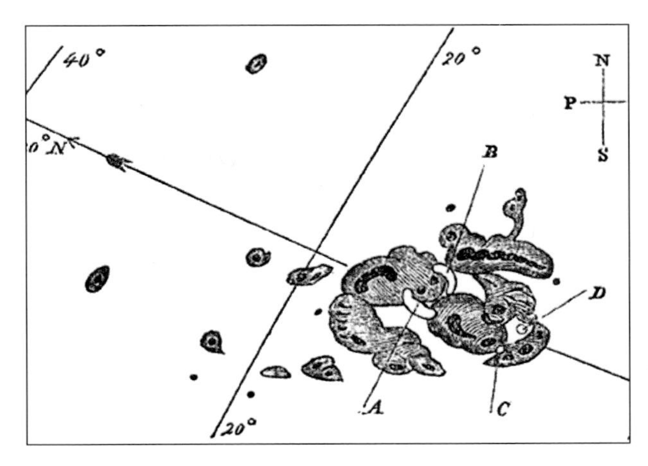

Drawings of the massive sunspot group and white light flare seen on the surface of the Sun on September 1, 1859.

As night fell on September 3, startled residents as far south as Puerto Rico were treated to something most of humanity rarely sees: a vibrant display of **aurora borealis**, or northern lights.

A great auroral display is one of the most spectacular natural sights to behold. I'd easily put the spectacle up there with a total solar eclipse and a really powerful meteor storm . . . and just maybe, a nearby galactic supernova or a bright daylight comet, just not *too* close!

What auroras are (and aren't): Like sunspots, legends of a strange nightly glow seen in the far north were almost mythical in ancient times. Once explorers struck out to higher latitudes, tales of northern lights became a reality, though residents of those areas toward the poles were always familiar with them and incorporated them into local lore. Even today, a sudden appearance of auroras toward lower southern latitudes can spark concerned calls to police and fire departments.

I've also heard some strange misconceptions from the general public as to what auroras are. One common misconception is that they're somehow the reflection of sunlight off of the polar ice caps (they aren't). I once heard a coworker in Alaska say that he thought auroras were more frequent on colder nights—there is a causal connection here, as cold winter nights also happen to be crystal clear—but they merely facilitate views of auroras, not the auroras themselves.

There was a suspicion that magnetic activity was linked to auroras even in Carrington's day. Anders Celsius and Olof Hiorter pioneered work linking auroras seen over Sweden in 1741 with terrestrial magnetism. Today, we know that highly charged particles on the solar wind interacting with the magnetosphere of the Earth cause auroras. These interact with the tenuous molecules of gas in the upper atmosphere, causing them to glow. Think of a neon tube, as the gas within glows a desired color as current is passed through the bulb. Though a greenish glow (due to ionized oxygen in a low energy state) is the most common hue seen, vibrant reds (nitrogen or high energy oxygen) and shimmering yellows (a red/green combination) may accompany more energetic displays.

A brilliant auroral display over Finland. DSLR w/14mm lens, f/2.8, 5-second exposure, ISO 1600.

WATCHING FOR AURORAS

Latitude is the key factor in seeing auroras. Ironically, the North Pole isn't the very best place to see aurora . . . the sweet spot for the auroral oval is right around latitude 60 to 65 degrees north, corresponding with Anchorage, Alaska; Whitehorse, Canada; Nuuk, Greenland; and the major capitals of Scandinavia. These regions actually enjoy auroras on most clear winter nights, though they vanish from view in the continual daylight of summer. Down under, there's a southern analog to the northern lights known as the **aurora australis**, or southern lights. These are tougher to catch, however, as southern continents do not extend as far southward in latitude. Tasmania, the southern island of New Zealand, Tierra del Fuego, and Antarctica are your best bets for catching the southern auroras.

If you're south of the 60, don't despair . . . bright auroras are visible toward lower latitudes around 45 degrees north (including the northern contiguous United States) on average about half a dozen days out of the year. I lived in Alaska near the Arctic Circle for 4 years, but the very best auroral display I ever witnessed was in northern Maine in 1982.

Many people see the typical dull-green halo of common auroras to the north and fail to realize what an amazing potential is possible. Really active auroras will display moving multicolored curtains and are truly eerie to watch. They're also bright enough to cast flitting shadows across the snow and make nighttime motorists stop to brave the chilly night air in wonder.

A fine example of a complex solar halo seen over Sweden, showing a halo, upper tangent arc, sundogs or "mock suns" on either side of the halo, and a parhelic circle connecting the two with the Sun.

Photographing auroras is also pretty straightforward: If a display is underway, simply take a series of wide-field exposures that are about 5 to 10 seconds long and see what turns up. Sometimes, auroras are too faint to be seen . . . but they will still show up on long time exposures of the night sky. Be sure to carry an extra set of camera batteries in a warm pocket or connect your camera to an external power supply, as time-exposure photography at sub-zero temperatures will drain batteries quickly. If the aurora is really strong and fast-moving, shorten the exposure time so the delicate structures don't become a blur.

Finally, keep an ear out for another curious phenomenon. For years, observers have reported hearing a distinct crackle or hiss accompanying auroras. Now, auroras occur about 50 miles (80 km) up high in the atmosphere, where sound *shouldn't* carry . . . what's going on here? Like audible meteors, skeptics long thought sounds accompanying auroras were simply a psychological phenomenon. However, in 2016, a team of Finnish researchers captured recordings of this eerie swishing sound during an auroral display.[15] This suspected culprit? A phenomenon known as **electrophonic sound**, where the high voltage current from an aurora sets up a localized energy field, resulting in an otherworldly crackling noise.

OTHER ATMOSPHERIC PHENOMENA

The atmosphere above can also give rise to some pretty bizarre sights. Halos often appear around the Sun and the Moon, as ice crystals from a high incoming front refract sunlight. Bright **sundogs** or **mock suns** can also punctuate a **solar halo**, a very dramatic sight. In addition to rainbows, there are also **moonbows**, **fogbows**, and more. These fascinating optical effects occur based on the shape, density, and orientation of those ice crystals, which are suspended high in the atmosphere.

Check out Les Crowley's Atmospheric Optics (https://www.atoptics.co.uk/) page for an exhaustive list of rare atmospheric phenomena.

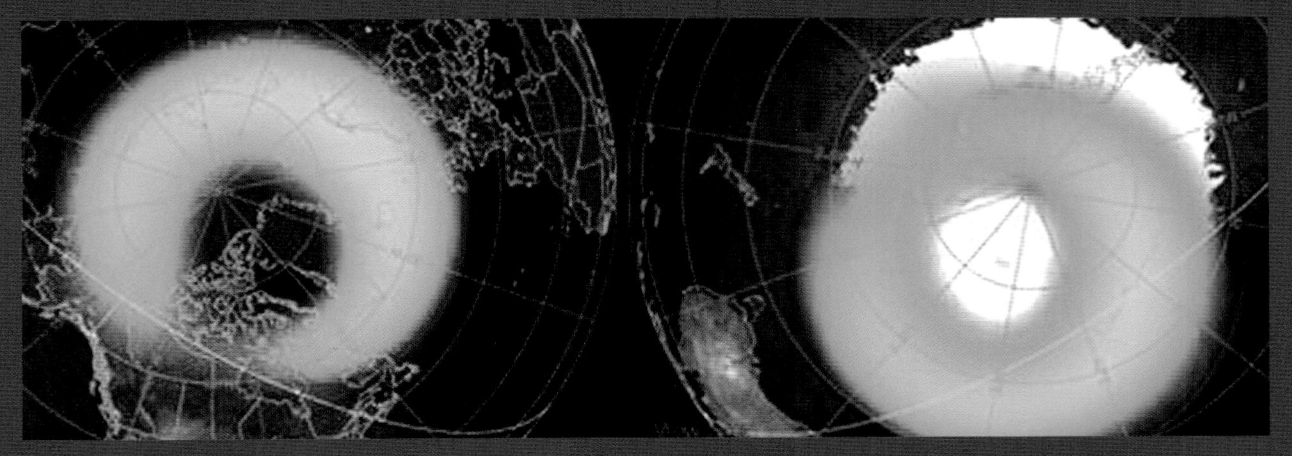

A diagram showing the typical position of the auroral oval over the northern and southern poles of Earth. The forecast also shows the position of the day/night terminator, and the visibility line from which auroras may be seen.

ACTIVITY: AURORA OBSERVING: WILL THERE BE AURORA TONIGHT?

Maybe you've wanted to see them your whole life, or you're a seasoned aurora-chaser wondering if you should bother setting up. Fear not. While it is indeed worth making that once-in-a-lifetime trip to the Arctic or Antarctica to see the aurora (just not in the summertime, when it's perpetual daylight!), you still have a good shot on seeing aurora if you live above 45 degrees north or south . . . if you know precisely when to look.

First, the aurora follow solar activity, which spikes and ebbs with an 11-year cycle. The last peak for the solar cycle was Cycle #24 on 2014, and the next peak is due around 2024–2025. The Sun seems to be a chronic underperformer as of late, making solar astronomers wonder if there aren't larger cycles of activity. For example, 2009 featured 260 spotless days—the most in over a century—and 2018 and 2019 are on track for a similar lack in activity. Keep in mind, however, that a massive coronal mass ejection can buck the overall trend and can always occur, even near solar minimum.

Also, for reasons still not fully understood, auroras seem to peak biannually right near the March and September equinoxes.

Generally, a large visible sunspot turned Earthward is a fertile breeding ground to launch a large solar flare or coronal mass ejection. The heavier energetic solar particles usually take 2 to 3 days to reach the local vicinity of the Earth, giving observers ample time to prepare. NASA's Advanced Composition Explorer (ACE) satellite usually sounds the first alarm that a space weather event is inbound. Once the alarm is sounded, watch the planetary K index, a measure of disturbance in the Earth's magnetic field. A Kp over 4 means that auroral activity is at storm levels, and 7 or higher means there's a good chance that the auroral oval is pushed down to lower latitudes.

How a coronal mass ejection impacts the magnetosphere of the Earth makes a big difference as to whether we're favored for bright auroras. This is known as the Bz value of an event. A plus value (with the Bz northward) is an unfavorable orientation, and a negative (with the Bz southward) is favorable. Think of magnets back in high school physics class, with the north and south poles of similar magnets repelling (Bz northward) and dissimilar poles attracting (Bz southward).

But in this case, the magnets are the Earth and the Sun. We inhabit the powerful heliosheath of our tempestuous host star.

Finally, it's *always* worth glancing northward for the dull-green glow of the aurora if skies are dark and clear. They can and do occasionally pop up without warning.

Social media (especially Twitter) is also a great resource. For example, folks in the United Kingdom and Scandinavia north of latitude 50 degrees frequently enjoy aurora displays. Located eastward of North America, they typically see the auroras first. If they're tweeting about the northern lights, there's a good chance that North American residents will see them come nightfall about 6 hours later. Simply search Twitter for words such as "aurora," "northern lights," and "space weather" to see what folks are talking about worldwide.

There are many apps available for your mobile device that'll notify you with customized alerts when aurora activity is high in your area.

Check out our list of handy aurora-observing resources on page 226.

COSMIC INTRUDERS: OBSERVING COMETS, ASTEROIDS, AND METEOR SHOWERS

How to track interlopers entering the inner Solar System.

"Year of meteors! Brooding year [. . .]
Year of comets and meteors transient and strange [. . .]
What am I but one of your meteors?"
—Walt Whitman, "Year of Meteors, (1859–60)"

NOTHING QUITE STIRS UP EXCITEMENT

in the astronomical community and the mind's eye of the general public like the sudden appearance of a bright comet.

ACTIVITIES

- Tracking comets in the sky online (page 129)
- How to observe and report meteor showers (page 139)

The inner coma of Comet Hyakutake, captured by the Hubble Space Telescope on April 3 and 4, 1996.

Left corner: A blue comet meets the blue sisters of the Pleiades: Comet C/2016 R2 PanSTARRS makes a close pass near the open star cluster Messier 45 on the night of February 3, 2018. A stack of 15x 5-minute exposures at ISO 1600, plus 15x Dark frames and 20x Bias frames.

It's hard to imagine today: Early on the morning of January 12, 1910, diamond miners in Transvaal, South Africa,[1] were just getting off the graveyard shift. Looking low to the east as dawn broke, these workers were the first to notice a great celestial spectacle about to unfold: a bright naked eye comet sat low in the dawn sky, already gracing the skies of Earth. Telegrams flashed the news around the world. The astronomer Robert Innes at the Transvaal Observatory in Johannesburg, South Africa, got his first good look at the comet on January 17. The public, already primed for the expected passage of Halley's Comet on that same year, marveled as the Great January Comet of 1910 raced northward.

Nothing quite stirs up excitement in the astronomical community and the mind's eye of the general public like the sudden appearance of a bright comet. My grandmother witnessed the Great January Comet of 1910 from Saint Francis in northern Maine, recounting how it was easily visible with the naked eye during the daytime. It seems that the nineteenth century was also awash with great comets . . . today, many modern star party attendees look at one of the half dozen-odd comets that are visible every year as fuzzy little starlike points in the eyepiece and wonder what the big fuss is all about. Though the next great comet could appear at any time, it always seems like we get bright naked eye comets in pairs; witness the Great January Comet of 1910 and Halley's Comet,[2] along with Hale-Bopp and Hyakutake back-to-back in 1996 and 1997.

A cometary nucleus up close: the European Space Agency's Rosetta mission gave us some amazing views of periodic Comet 67/P Churyumov-Gerasimenko right up until the end of its mission on September 30, 2016.

What are comets? Comets are dirty snowballs, chunks of water ice and dust. Astronomers believe a great reservoir of comets exists in the **Oort Cloud** in the far outer Solar System from 5,000 AU to 200,000 AU from the Sun. To give you some idea of just how far out that is, the orbit of Pluto ranges from 30 to 49 AU from the Sun, and the Voyager 1 spacecraft—the most distant spacecraft from Earth—is, as of December 7, 2017, 140.5 AU from the Sun. The outer Oort Cloud is an appreciable fraction of the distance to the next star—about 75 percent of the way to Proxima Centauri—and the occasional close passage of a star in the far distant past may send a comet on its millennia-long journey toward the Sun. Some comets may get captured in shorter period orbits and become frequent visitors to the inner Solar System. First-time comets entering the inner Solar System stand roughly a 40-percent chance of having Jupiter's gravitational influence alter their orbit.[3] Early astronomers considered these hairy stars as an atmospheric phenomenon, and nearly every astrologer of the day considered their appearance as an ill omen. One apocryphal tale has Pope Calixtus III excommunicating Halley's Comet for its appearance 3 years after the fall of Constantinople to the Ottoman Turks on May 29, 1453. Astrologers also blamed the 1665 outbreak of the plague on another bright comet, seen the winter before. Shakespeare gave comets more bad PR in his play *Julius Caesar*:

"When beggars die, there are no comets seen; the heavens themselves blaze forth the death of princes."

Today we know that far from their medieval role as meddlers in terrestrial affairs, comets are much more interesting, and possibly more intimately connected to life on Earth than we could have imagined. Not only are they a suspect source for delivering the water in Earth's oceans inside the frost line of the inner Solar System, but they may have delivered the first organic molecules to the primitive Earth. The European Space Agency's Rosetta mission to Comet 67/P Churyumov-Gerasimenko in 2014 not only delivered the Philae lander to the first rough and tumble landing on a comet, but also fundamentally changed the way science thinks about comets as primordial building blocks in general.

NAMING COMETS

Here's the good news: Discover a comet, and you have a minor piece of immortality. The International Astronomical Union (IAU) designates new comets by the year, time of year discovered, and the name of the discoverer, amateur, or professional. The IAU will also accept up to three names if there are simultaneous independent discoveries worldwide in a 24-hour period . . . that's how we ended up with tongue-twisting names such as Comet C/1983 H1 IRAS-Araki-Alcock and 45/P Honda-Mrkos-Pajdušáková. Astronomers denote the time of year of discovery and the order when the comet was announced using the letter/number suffix in the middle of a comet's designation. The system divides each month in half, with "A" as January 1 to 15, "B" as January 16 to 31, and so on.[4] For example, Comet C/2017 S3 PanSTARRS was the third comet discovered in the first half of July 2017.

The rise of automated astronomical surveys means an increasing number of robotic eyes are now discovering new comets as they routinely scour large swaths of the sky every night down to ever-fainter magnitudes. We're seeing lots more comets with names of such surveys as PanSTARRS or LINEAR these days.

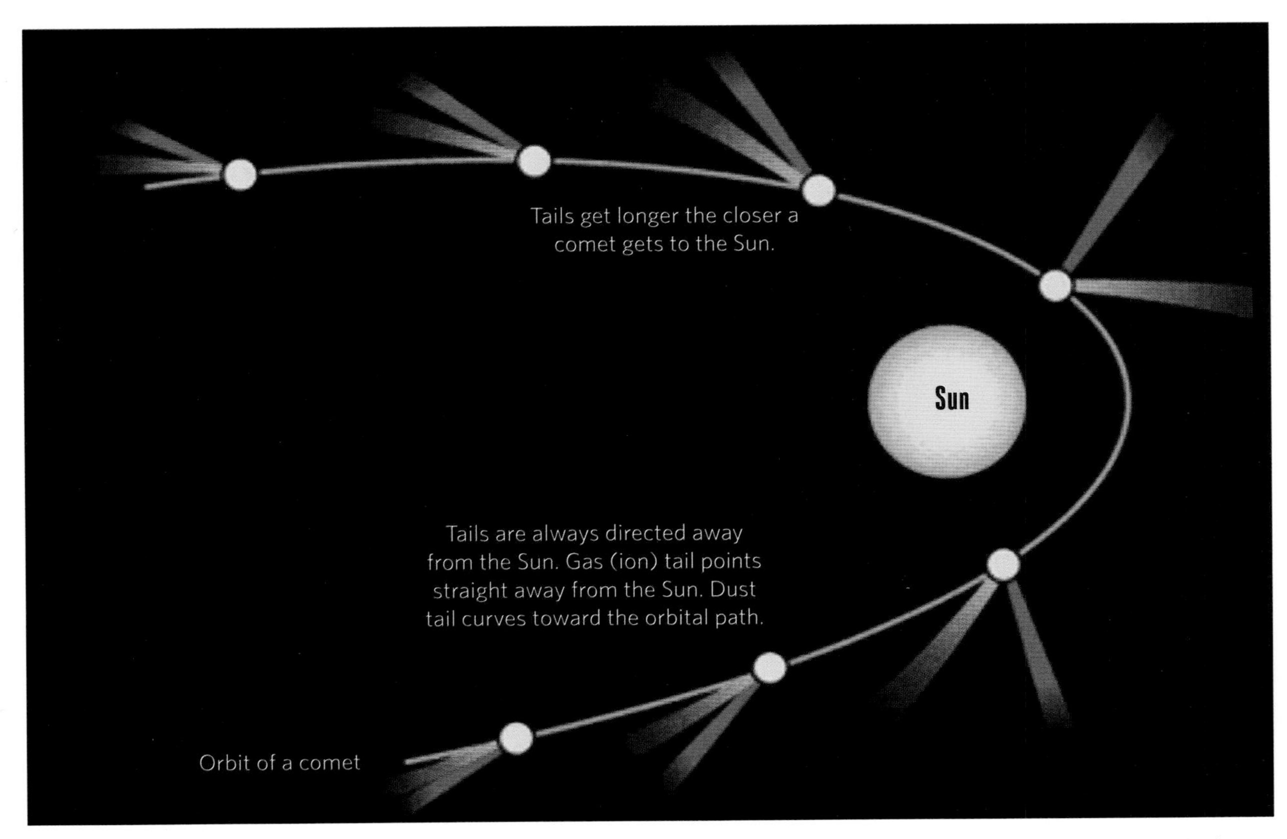

The path of a long period comet through the inner Solar System. As the comet nears perihelion, solar heating and solar wind typically pushes back a dust (yellow) and ion (blue) tail. These tails always extend away from the Sun, even when the comet is headed back out of the Solar System, tail first.

TYPES OF COMETS

Of the 3,996 comets thus far discovered (as of December 2017), 363 have **periodic orbits** shorter than 200 years. **Periodic comets** go all the way back to the most famous comet of them all, 1P/Halley. The astronomer Edmond Halley calculated the orbit of the comets seen on previous apparitions in the years 1607 and 1682, and discerned that they were actually one comet on a 75.3-year orbit. Halley correctly predicted the comet's return in 1758. Though Halley never lived to see his famous comet return, astronomers picked it up just before the year was out on Christmas Day, 1758.

A vast majority of comets are on long-period orbits measured in thousands of years or more, often making their first appearance in the inner Solar System. Some comets are on near-hyperbolic open orbits with an eccentricity (how far its orbit deviates from circular) very near 1.0. Astronomers long suspected that some of these comets may actually be interstellar visitors to our Solar System, but the very first object with a proven eccentricity greater than 1.0 was actually an asteroid: the strange elongated body 1I/2017 'Oumuamua, which raced through the inner Solar System in late October 2017.

Most long period comets have distant aphelia, and never enter the inner Solar System. What really grabs astronomers' attention is a new long period comet entering the inner Solar System discovered while it's still far out, suggesting it's a large and intrinsically bright object. Though the dusty coma of a comet can be larger than a planet with a dust tail spanning millions of kilometers, the icy nuclei of a comet is tiny, generally only 10 miles (16 km) across. **Sungrazing comets** (or **sungrazers**) may approach so close to the Sun that they disintegrate entirely. Three well-studied classes of sungrazers are the **Kreutz**, **Marsden**, and **Kracht** family of sungrazing comets. We know of 3,398 sungrazing comets (as of December 7, 2017), thanks to the sunward-staring Solar Heliospheric Observatory (SOHO) space telescope. Prior to its launch in 1995, astronomers only knew of less than a dozen confirmed sungrazing comets.[5]

The brilliant, sungrazing Comet C/2011 W3 Lovejoy as imaged from the International Space Station, post-perihelion.

COMETS: WHAT TO EXPECT AT YOUR EYEPIECE

Binoculars are great for sweeping up comets down to about +10th magnitude or brighter. A larger light bucket telescope will extend this view down to about +13th magnitude. Comets often start out as little fuzzy stars, appearing like small, indistinct globular clusters whose stars stubbornly refuse to snap into focus and resolution. Watch a comet for a few hours, and you might notice its movement against the starry background, a sure sign of a cosmic interloper.

I wonder how many comets fainter than naked eye visibility came and went undocumented before the era of the telescope, just a few short centuries ago?

And like globulars and nebulae, that quoted magnitude of a comet is smeared out over its apparent angular surface area at the eyepiece, making it appear fainter than you'd expect. This is why a +10th magnitude comet is often still a tougher find than a star of the same brightness. Make sure you're viewing from as dark a sky site as possible, and light pollution or an OIII filter can help tease a comet out under washed out, low-contrast skies.

How do comets progress and develop as they enter the inner Solar System? The 1996–1997 passages of Hale-Bopp versus Hyakutake gave us a good study in two types of bright comet apparitions: Hale-Bopp was a large, bright comet anticipated for years and seen from a distance, while Hyakutake was a smaller comet seen close up. While the astronomical community was eagerly awaiting Hale-Bopp, Hyakutake snuck up and surprised us all with an amazing preshow months prior. Had Hale-Bopp appeared 6 months earlier or later, we would have been in for a truly amazing spectacle. As it was, Hale-Bopp lingered as an amazing sight for northern hemisphere observers for several months.

Comets are often fickle, and sometimes over- or under-perform versus expectations. Comet Kohoutek, for example, was a famous fizzle back in the 1970s. More recently, Comet ISON failed to hold it together post-perihelion on U.S. Thanksgiving Day 2013, another comet fail. On the other side, comets are also prone to great outbursts. Comet 17/P Holmes, for example, flared into naked eye view in October 2007, while it was still over 2 AU from the Sun. Comet C/2011 W3 Lovejoy passed 87,000 miles (140,000 km) from the surface of the Sun on December 16, 2011—closer than the Earth-Moon distance, and closer than the perihelion passage that doomed Comet ISON—and emerged intact from behind the Sun and went on to become a great dawn comet for southern hemisphere observers.

To quote the veteran comet hunter David Levy, "Comets are like cats: they have tails, and they do exactly what they want."

TALES OF COMET TAILS

The next feature you might notice as a comet develops is a short, spiky tail. This can progress into the magnificent fan of a **dust tail** as the comet nears the Sun. The solar wind sweeps the tail back in the anti-sunward direction,[6] and the comet may develop a second luminous twin **ion tail**. These two tails might overlap or appear nearly opposite to one another, depending on the approach geometry. Sometimes, a comet might also display a short, spiky **anti-tail** in the sunward direction.

We say a comet "goes green" when it starts to show a subtle emerald color due to ionized cyanogen gas.

Some great comets of the past century are listed in the table on the next page.

TOP COMETS OF THE PAST CENTURY			
Name	**Perihelion**	**Magnitude**	**Notes**
C/2011 W3 Lovejoy	December 16, 2011	-5	Passed 87,000 miles (140,000 km) from the Sun
C/2006 P1 McNaught	January 12, 2007	-5.5	Brightest in twenty-first century
C/1995 O1 Hale-Bopp	April 1, 1997	-0.5	Discovered 7.2 AU from the Sun
C/1996 B2 Hyakutake	May 1, 1996	0	Passed 0.1 AU from the Earth
C/1975 V1 West	February 25, 1976	-3	Followed Kohoutek
C/1969 Y1 Bennett	March 20, 1970	0	Was to be imaged by Apollo 13
C/1965 S1 Ikeya-Seki	October 21, 1965	-10	One of the brightest comets
C/1962 C1 Seki-Lines	April 1, 1962	-1	
C/1956 R1 Arend-Roland	April 8, 1957	-1	
C/1948 V1 "Eclipse Comet"	October 27, 1948	-1	Discovered during a solar eclipse
Great Southern Comet	December 4, 1947	0	Faded soon after discovery

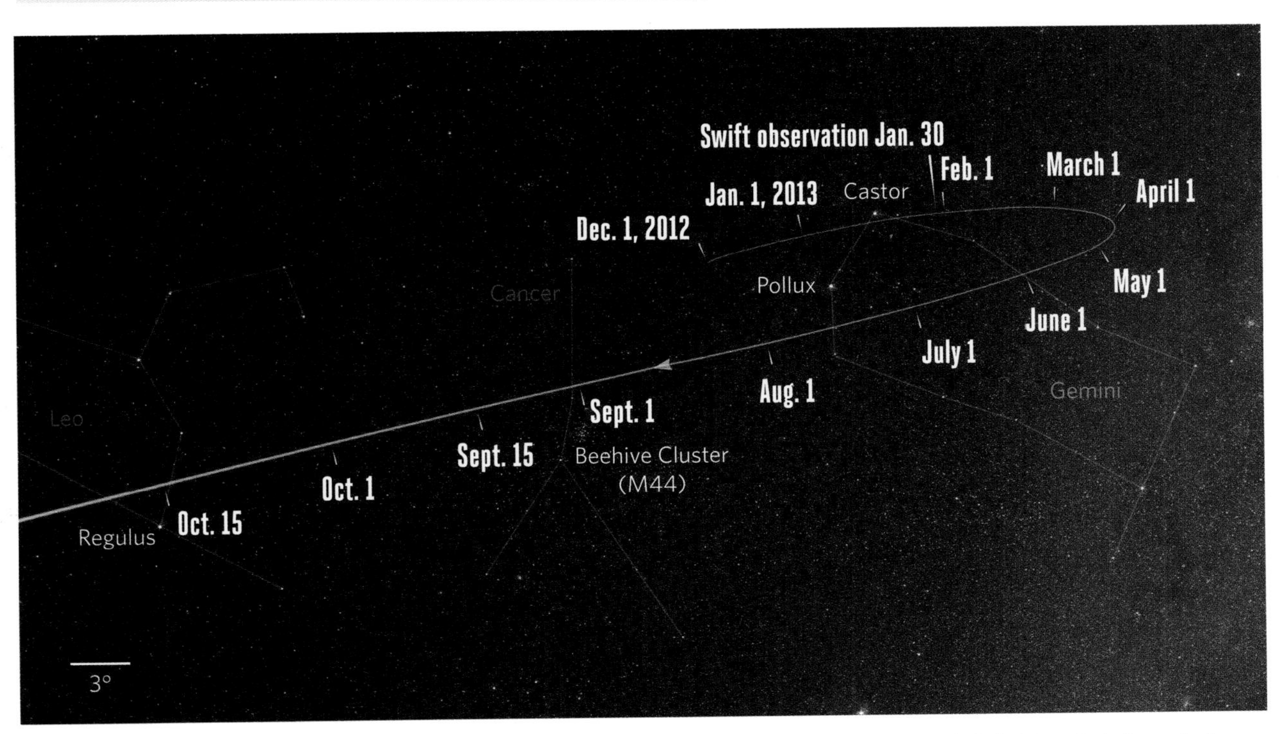

The typical looping path of a comet through the sky—in this case, doomed sungrazing Comet C/2012 S1 ISON, which distintegrated after perihelion on U.S. Thanksgiving Day, 2013.

Comet 2014 Q2 Lovejoy from January 13, 2015. Note the green color due to cyanogen gas. 200 mm lens and f4, 3200 ISO.

TRACKING THE MOTION OF COMETS AND IMAGING

As comets start to break naked eye magnitude (+6) they start to become bright enough to image with foreground objects. The good news is this type of photography is relatively easy and straightforward, similar to shooting star trails: A simple wide-field 10- to 15-second exposure with a DSLR camera on a sturdy tripod should reveal the fuzz ball of a bright comet against the starry background. You'll need to track longer exposures of a minute or more to avoid star trails. You can easily track a comet using a computerized mount or our low-tech barn door tracker (see Chapter 9: Astrophotography 101, page 150).

HOW TO REPORT A (POTENTIAL) COMET DISCOVERY

Scenario: It's a cool autumn night. The Moon is absent from the sky, and the stars shine bright and steady with nary a twinkle. It's the kind of night that you just *know* it's a crime to pass up. Winter is on its way, after all, and there are plenty of cloudy nights when your telescope will sit dormant. You set up, then begin patiently sweeping through the sky at low power. Maybe you have a target such as a bright nebula or a globular cluster in mind, or perhaps you are purposely on the hunt for something a bit more . . . unexpected.

Suddenly, a small fuzzy patch snaps into view. Sweeping back, you tweak the focus one way, then the other, as the fuzzy patch stubbornly refuses to snap into focus. You check the object's position in right ascension and declination, then refer to your handy red-lit star charts for any known nebulae in that position. Zip. You look back at the fuzzy blob . . . has it *moved* slightly in the last few minutes? It's hard to tell. You realize you might've just nudged the telescope a bit in the excitement of the moment. Though the rational part of your brain tells you there are a hundred reasons it might not be so, a small voice also tells you something truly exciting: The fuzzy smudge might just be a new comet.

What now?

Well first, the bad news. You've got lots of competition out there, both human and robotic. Once almost solely the domain of dedicated amateur observers, surveys such as LINEAR, PanSTARRS, and more are increasingly finding comets at even fainter magnitudes. Still, amateur finds do occasionally occur. The advanced amateur astronomers Alan Hale and Thomas Bopp actually discovered Comet C/1995 O1 Hale-Bopp separately while casually scanning the region around the globular cluster Messier 70. The australian astronomer Terry Lovejoy discovered six comets starting with Comet C/2007 E2 Lovejoy in 2007. Of course, he has the home court advantage of an enviable southern hemisphere vantage point, with much less competition.

It's one of the few assured avenues to astronomical immortality for an amateur. Be the first to discover and report a comet, and it will bear your name and up to two other names. For some reason, astronomers with umlaut-laden names seem to always discover comets.

ACTIVITY: TRACKING COMETS IN THE SKY ONLINE

Did you know you can actually discover comets on the web?

One great resource is SOHO's LASCO C3 camera. Launched in 1995, SOHO continually stares at the Sun, monitoring solar weather around the clock. With a 15-degree-wide field of view, SOHO also picks up lots of sungrazing comets that would otherwise pass through the inner Solar System undetected. To date, dedicated amateur sleuths have discovered nearly 4,000 sungrazing comets. This is done entirely using SOHO LASCO C3 data, which is freely available online. The process is relatively easy, though remember, you've got lots of competition scouring those same images worldwide.

SOHO posts short animated GIFs for the last 24 hours at their main site: https://sohowww.nascom.nasa.gov/

Visually examine these with the highest resolution your monitor can bear. You will see lots of hash, as cosmic rays and high energy particles on the solar wind send sparkles across SOHO's detectors. You will also see planets and stars moving across the field of view.

Here's a handy link for known transits of planets, asteroids, and comets across SOHO's LASCO C2/C3 field of view: https://sungrazer.nrl.navy.mil/index.php?p=transits/transits

A comet will trace out a path nearly directly toward the Sun, which is hidden behind the center of the occultation disk. It's useful to watch some SOHO sungrazer videos first, to get a good feel for what a sungrazer looks like. SOHO also has a handy page for the projected paths for different sungrazer family groups of comets at different times of the year: https://sungrazer.nrl.navy.mil/index.php?p=comet_tracks

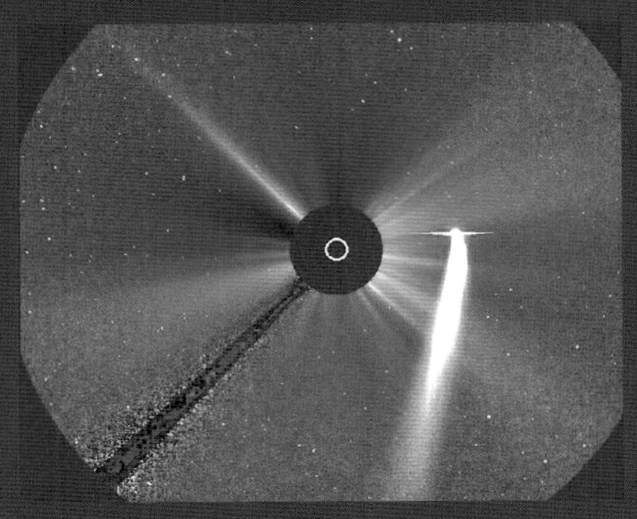

Sungrazing Comet C/2004 F4 Bradfield passes near the Sun and through the field of view of the Solar Heliospheric Observatory's (SOHO) LASCO C3 camera.

You've seen a suspect comet: What's next? Well, first you should check to see if it's a known comet. Karl Battams maintains a handy web page for SOHO sungrazers, including discovery discussions and reporting procedures: https://sungrazer.nrl.navy.mil/

Sleuthing out sungrazers is a great cloudy night activity, and hey, you're doing real science to boot. Good luck!

You can only hope that the newly found comet bearing your moniker is merely a "great comet of the century," rather than a once-every-100-million-year extinction-level event. The rule seems to be that for every naked eye comet, there are perhaps ten comets that reach binocular visibility (+10th magnitude) or brighter, then maybe ten more that fail to achieve even that. Your comet will receive a name such as "Comet C/2017 A1 Smith" where the C or P at the front denotes P for Periodic with an orbit around the Sun of less than 200 years or (more likely, since most of the large periodic comets in the Solar System have long been discovered) a C for a long-term orbit of a temporary visitor to the inner Solar System from the distant Oort Cloud, perhaps for the very first time. Comet C/1996 B2 Hyakutake, for example, wowed astronomers in 1996 and won't return for another 70,000-odd years.

Comet discoveries are reported to the International Astronomical Union's Central Bureau for Astronomical Telegrams, or CBAT.[7] CBAT suggests that, in addition to watching for motion, the observer check this discovery against a list of known comets and record the position down to 1-arcminute declination and 0.1-arcminute right ascension against planetarium software or a good star atlas. Consulting a professional observatory (if possible) is also a good idea, though I'd be leery about letting the cometary cat out of the bag online before CBAT confirms the discovery with backup observations worldwide; comet hunters are a pretty competitive bunch, and you don't want to get scooped!

Finally, perseverance is key when it comes to comet hunting, and you should always keep a vigilant eye out for something amiss, even during casual observations. Veteran comet hunter David Levy swept the skies for over a decade before finding his first comet, Comet Levy-Rudenko, in 1984, and went on to codiscover the historic Comet Shoemaker-Levy 9, which crashed into Jupiter in July 1994.[8]

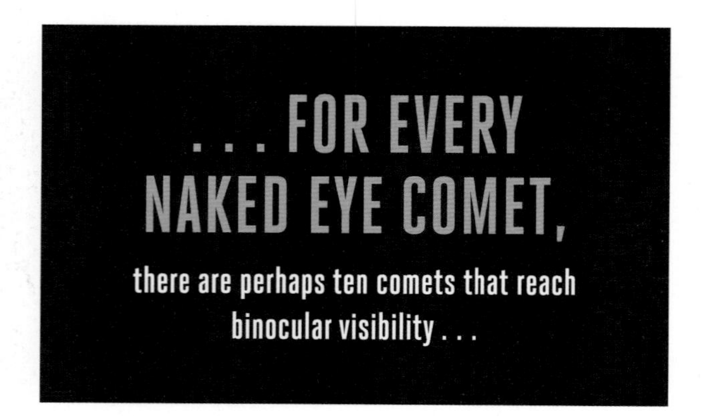

OBSERVING METEOR SHOWERS

It's still one of the coolest things we've ever seen.

As luck would have it, I found myself in late 1998 living in a tent in the deserts of Kuwait. I was deployed to al Jaber Air Base with my U.S. Air Force squadron from Eielson Air Force Base, Alaska. Unlike military installations stateside, forward operating bases are dark places at night. I had been watching the annual Leonid meteor shower for the past few years, but nothing prepared me for the spectacle that unfolded near dawn on the morning of November 17, as the zenithal hourly rate approached 1,000, with bright fireballs punctuating the night and lighting up the desert floor every few seconds.

This was no paltry meteor shower, spitting out a mere faint meteor every 10 minutes or so. Every 33 years, the November Leonids put on a spectacular display, shooting thousands of meteors per hour from the Sickle asterism in the constellation Leo. Such a spectacle is rarer than a total solar eclipse. And the good news is, while a **meteor storm**—a shower with an observed rate of 500 meteors per hour or more—is rare, there are about a dozen **major meteor showers** topping 20 or more meteors per hour every year, and you don't need special gear to watch them, just clear skies and patience.

What are meteors? First, let's clear up a common misconception. Many folks see giant space rocks in a museum and assume that's what all meteors are. Some folks even think that meteors and meteor showers are dangerous. Truth is, a majority of what you're seeing during a meteor shower are tiny grains of dust. Comets lay these dust trails down, and meteor showers occur where these ancient paths intersect the Earth's orbit. Occasionally, a *really* bright fireball might result from a pea-size rock. To date, no one has recovered a meteorite related to an annual meteor shower.

A composite image of the 2017 Geminid meteor shower over Helgeland, Norway.

METEOR SHOWER BASICS

The study and observation of meteor showers has a lingo all of its own.

SPORADICS

These are meteors that aren't related to known showers. Meteor showers are named after the constellation they appear to radiate from. This point in the sky, known as the **radiant**, is an optical illusion, similar to how a long highway seems to converge to a point on the distant horizon. For example, the August Perseids[9] seem to hail from the constellation Perseus. Trace a meteor back to elsewhere in the sky in mid-August, and it's very probably a sporadic. The average background sporadic rate varies from around two to five meteors per hour, throughout the year.

ZENITHAL HOURLY RATE (ZHR)

This is the number of meteors from a given shower that you can expect to see, under optimal conditions. Keep in mind, this is an *ideal* rate, assuming the radiant is overhead and the sky is completely dark. The true position of the radiant, light pollution (including the phase of the Moon), and observer limitations (you can't see the entire sky at once) all conspire to make the true observed rate lower than the ideal ZHR.

METEOR SHOWERS

are named after the constellation they appear to radiate from.

VELOCITY

The speed of the meteor you see sliding through the sky tonight is the product of two velocities: the Earth, barreling around the Sun at 19 miles (30 km) per second, and the meteor's velocity, as it hits the Earth either head-on, adding its final impact velocity to our own in a big cosmic splat, or at an oblique and slower angle from the side or behind. Ever driven at night in a snowstorm? Then you've seen a similar effect looking straight ahead, as snow shining in your high beams seems to all converge from one point, straight in front of the car. There's another useful car analogy when it comes to meteors: The forward grill and windshield gets all the bugs. The same is true for the Earth versus these ancient debris trails that comets laid along its path around the Sun, long ago. The Earth carves out a tunnel 8,000 miles (12,880 km) wide and 1.6 million miles (2.6 million km) long, every day. Past local midnight, you're standing on the half of the planet turned forward into the Earth's path, and the meteor shower rates pick up with swift, head-on meteors. That's why the very best time to watch a particular shower is in the zero-dark, early pre-dawn hours. Any occasional straggling meteors you see in the early evening need to catch up to the Earth, leaving slower passages through the sky.

FIREBALL

Once in a while, a meteor brighter than -4th magnitude (equal to Venus at its brightest) occurs, perhaps casting a shadow during its passage. A fireball with a bright terminal explosion flash at its end[10] is sometimes referred to as a **bolide**. Keep those binoculars handy, as a fireball can leave a lingering glowing train and smoke trail, which may persist for several minutes.

Very rarely, you might just witness what's known as a **meteor procession**, or the long stately trail of a major meteor splintering on atmospheric entry, perhaps on its way to becoming a meteorite. The American poet Walt Whitman witnessed just such a display in 1860, as recounted[11] in his poem "Year of Meteors, (1859–60)," which states,

". . . the strange huge meteor procession, dazzling and clear, shooting over our heads . . ."

Auroras, luminous high-altitude noctilucent clouds, and a brilliant Perseid fireball from 2016.

HOW METEOR STREAMS EVOLVE

The annual showers we currently see aren't permanent. Not only are these cometary debris streams clumpy or knotty, but they also evolve over time. Over the centuries, new showers move into the path of the Earth, while older streams drift away, mainly due to the pressure of the solar wind. The Geminids, for example, are now overtaking the August Perseids as the top annual draw, while the obscure Andromedids—a major nineteenth-century meteor shower—are also seemingly making a comeback. Meanwhile, the January Quandratids now display a narrow peak, with observers at different longitudes often missing the early January show entirely.

A painting depicting the Great Meteor Procession of July 20, 1860.

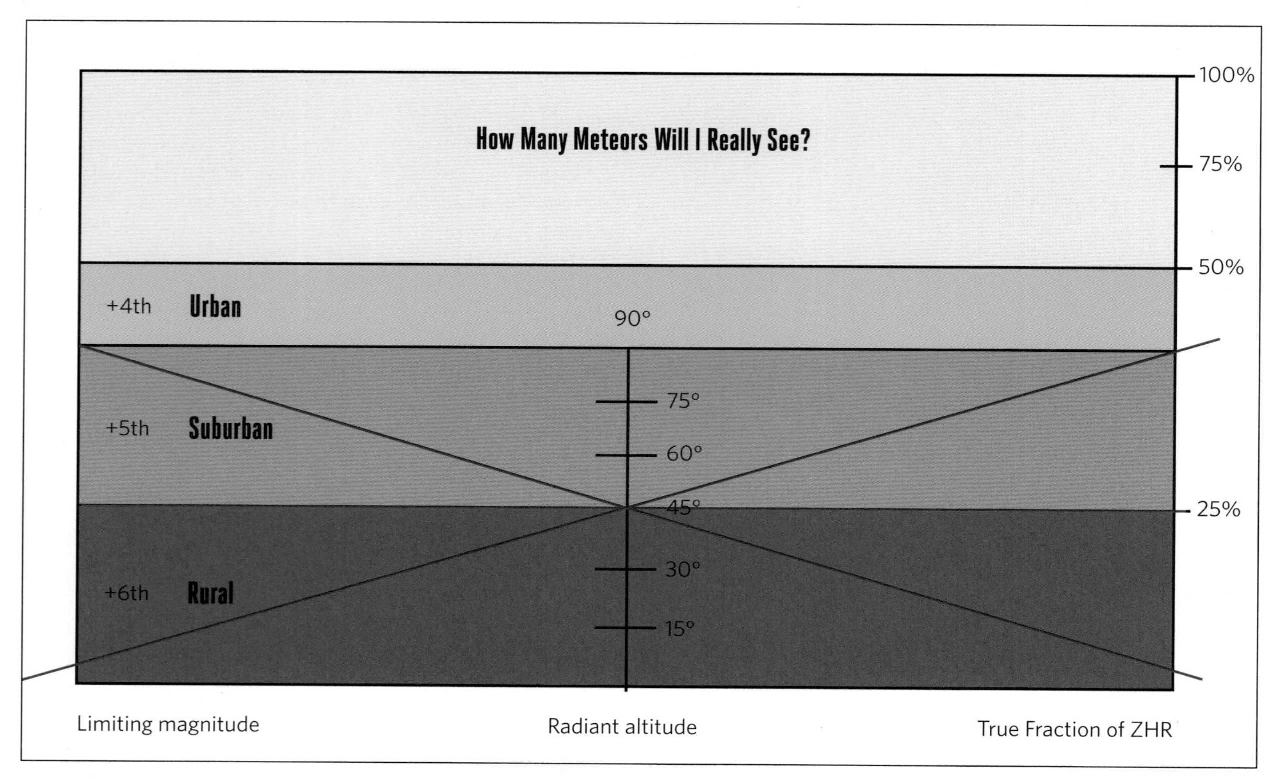

Light pollution, limiting magnitude, and the elevation of the meteor shower radiant all make a huge impact on the ideal Zenithal Hourly Rate (ZHR) and how many meteors you will really see. As shown in the nomograph above, even a slight drop in limiting magnitude coupled with a lower elevation for the radiant—most of us do not have the luxury of having the radiant directly overhead—makes for a precipitous drop in the hourly meteor count.

OBSERVING METEOR SHOWERS

Word of advice: Keep the telescope inside for a meteor shower. In fact, to observe meteor showers, you just need clear skies, a good set of working Mark-1 eyeballs, and patience. Do try and find as dark a site as possible, as light pollution dramatically cuts down on the number of meteors you'll see.

This classic nomograph calculator above highlights just how sensitive the crucial ZHR for a prospective meteor shower is versus the limiting magnitude and the elevation of the shower's radiant.

Enlist a friend (someone who'll get up with you at 3 a.m. and sit out in the cold and dark with you is a good friend, indeed!), as two sets of eyes looking in different directions will see more meteors than one. Dress warmly; I remember, growing up in northern Maine, that even August can be chilly at 4 a.m. Use a cot or reclining lawn chair to watch in comfort, but maybe not so comfortable that you fall asleep. If the Moon is up, try and position a building or a hill between it and your observing site to block out its glare. Draw a line from the limiting magnitude across the radiant altitude to reveal how many meteors (8 percent) you'll really see.

Also, keep a pair of binoculars handy, to check out lingering meteor fireball trails.

Which way to look? There are different schools of thought on this. Many seasoned observers say to look 45 degrees away from the radiant to catch oncoming meteors in profile. We like to simply vary our direction of view every hour or so, just to keep things interesting. Meteors can appear anywhere in the sky.

My personal rule: If skies are clear, I step out on the morning of a predicted meteor shower's peak, and just watch for 10 minutes. If I see one or more, I keep watching. If I see nothing after 10 minutes, I move on with my morning . . . or go back to bed. Sorry, meteor shower, you had your chance to thrill me. Do better.

A typical, lightweight DSLR rig complete with sky tracking mount and guide scope, suitable for wide-angle meteor photography.

IMAGING METEOR SHOWERS

DSLR cameras have simplified meteor photography considerably. As with satellites, simply taking tripod-mounted, wide-angle time exposure shots of the sky will catch anything that flashes past, including a bright meteor. We advise taking several test shots first, to get the combination of the f/ratio/shutter speed and ISO settings just right for the current sky conditions. Then you can automate the process using an intervalometer to take continuous timed shots. I like to take 30-second exposures, as our DSLR takes time to recycle between shots. Make sure you set the focus on a bright star or planet early on, or you'll end up with a string of out-of-focus shots. Also, watch for dew forming on the camera lens (see Chapter 2, page 43, on dealing with dew buildup).

Finally, start with freshly charged camera batteries, and keep a spare swap-out set in a warm pocket, or run the camera off of an external AC plug-in or external power pack; time exposure shots combined with cold morning temperatures can drain camera batteries in a hurry.

Review those images carefully. Nearly every meteor we've caught on camera wasn't apparent while we were watching, a testament to how many actually slip past!

METEORS ON THE MOON

When the Earth is in a meteor stream, so is the nearby Moon. If the Moon is at waning crescent phase past Last Quarter, it's worth running video aimed at the night side of the Moon near the peak of a meteor shower to see what turns up. Amateurs and automated programs have caught flashes on the unlit nighttime side of the Moon. One program named Lunar Scan (http://www.lunarimpacts.com/lunarimpacts_howto.htm) will even automatically comb through a video, flagging suspect flashes. NASA maintains a page (https://www.nasa.gov/centers/marshall/news/lunar/overview.html) for upcoming campaigns to watch for meteor shower flashes on the Moon during strong meteor showers.

BIZARRE METEORS

Often, observers report seeing green or yellow in meteors, signs of nickel or sodium content, respectively. One Japanese company, Astro Live Experience, or ALE,[12] wants to field multi-hued, artificial meteor showers starting in 2020, using small CubeSats to dispense tiny pellets. Also watch for corkscrew meteors that seem to refuse to follow a straight path. These physically *shouldn't* exist, though observers occasionally report seeing such oddities. Is this due to atmospheric ablation prior to the dark-flight of a meteor headed toward the ground, or is it merely an optical illusion?

Also, keep an ear out for a hiss or crackle accompanying some of the very brightest meteors. Now, meteors occur about 50 miles (80 km) up in the atmosphere, and sound (usually instantaneous) shouldn't carry to the observer on the ground . . . what's going on here? As with audible auroras (see Chapter 6: Near-Sky Wonders, page 104), acoustic meteors are a reality, due to a phenomenon known as electrophonic sound. A Japanese team caught meteors on audio in 1988 during the intense Perseid shower.[13] We've heard this spooky action from a bright Perseid before, though strangely, the intense 1998 Leonids were silent. (Perhaps the surrounding desert sand wasn't such a good electrophonic reflector?)

RADIO METEORS

Did you know you can also actually hear meteors crackle on the FM dial? A similar phenomenon occurs during a lightning storm. Most drivers are familiar with a popping crackle on the radio accompanying a flash of lightning while driving through a thunderstorm. In the case of meteors, their brief passage through the upper atmosphere ionizes a trail of hot gas, with an accompanying radio burst. Sometimes this can even act as a reflector, briefly bringing a distant radio station into focus. FM seems to work best; simply set the dial to an unused station low on the dial,[14] then listen to the static of the Universe as meteors slide silently past.[15]

THE BEST ANNUAL METEOR SHOWERS

TOP THIRTEEN ANNUAL METEOR SHOWERS					
Shower	**Radiant**	**Date**	**ZHR**	**Source**	**Notes**
Quadrantids	R.A. 15H 28′ Dec. +50°	January 3	60–200	2003 EH1	Narrow peak
Lyrids	R.A. 18H 8′ Dec. +32°	April 22	20	C/1861 G1 Thatcher	Rare outbursts
Eta Aquarids	R.A. 22H 20′ Dec. -1°	May 6	50	Halley's Comet	Complex stream
Daytime Arietids	In Aries	June 7	30	1566 Icarus	Radio shower
June Boötids	R.A. 14H 55′ Dec. +48°	June 27	50–100	7P/Pons-Winnecke	Variable
Delta Aquarids	R.A. 22H 35′ Dec. -16°	July 30	30	Unnamed Kreutz sungrazer	Rare southern shower
Perseids	R.A. 3H 4′ Dec. +58°	August 12	100–200	109P/Swift-Tuttle	A summer favorite
Taurids	R.A. 6H 24′ Dec. +15°	October 10	10–20	2P/Encke	High ratio of fireballs
Orionids	R.A. 6H 24′ Dec. +15°	October 21	20	Also Halley's Comet	Broad peak
Leonids	R.A. 10H 5′ Dec. +22°	November 17	20–10,000+	55P/Tempel-Tuttle	Strong outburst every 33 years
Andromedids	R.A. 1H 36′ Dec. +37°	December 3	10–50	3D/Biela	Making twenty-first century comeback?
Geminids	R.A. 7H 26′ Dec. +33°	December 14	120	3200 Phaethon	Stronger returns in recent years
Ursids	R.A. 14H 28′ Dec. +76°	December 22	10–70	8P/Tuttle	Making a comeback?

BEST YEARS 2019–2024

Note: (Moon is from Last to First Quarter)

Quadrantids: 2019, 2022

Lyrids: 2020, 2023

Eta Aquarids: 2019, 2021, 2022, 2024

Daytime Arietids: 2019, 2021, 2024

June Boötids: 2019, 2020, 2022

Delta Aquarids: 2019, 2022, 2024

Perseids: 2020, 2021, 2023

Taurids: 2020, 2021, 2023, 2024

Orionids: 2019, 2020, 2022, 2023

Leonids: 2020, 2022, 2023

Andromedids: 2019, 2021, 2024

Geminids: 2020, 2023

Ursids: 2019, 2022

IN THE EARLY NINETEENTH CENTURY,

asteroids were considered planets.

ASTEROIDS

As with comets, several dozen asteroids come within view of binoculars at +10th magnitude or brighter. Giuseppe Piazzi discovered the very first asteroid 1 Ceres on the first day of the nineteenth century, January 1, 1801.

In the early nineteenth century, asteroids were considered planets. As their ranks grew, however, astronomers soon realized these tiny worldlets not only seemed to frequent the space between Mars and Jupiter, but merited a category all their own.

Asteroids are designated in order of discovery and are named after nearly anything imaginable, from gods and goddesses to cities, rock stars, and even astronomy writers (though there isn't an asteroid named David Dickinson, at least not yet!).

Though no asteroid is large enough or approaches the Earth close enough to show a disk at the telescope, you can note their movement against the starry background by sketching or photographing the suspect field from one night to the next, and watching for movement.

Several times a year, a small asteroid will pass us closer than the Earth-Moon distance. Most of these are found shortly before closest approach by automated surveys, scouring the sky for inbound space rocks. On the night of October 5-6, 2008, the 13-foot (4.1-m)-wide asteroid 2008 TC3, discovered by the Catalina Sky Survey, generated considerable excitement when astronomers realized it would strike the Nubian Desert in Sudan just 19 hours later. 2008 TC3 went on to become the first meteorite recovered from a previously observed inbound asteroid. Others, such as J002E3, were later revealed to be space junk from previous missions such as Apollo 12, lazily chasing the Earth in solar orbit.

On the next page, you will find the best asteroids to track.

TEN BRIGHTEST ASTEROIDS				
Name	Orbital Period	Max. Brightness	Inclination	Next Opposition
1 Ceres	4.6 years	+6.7	10.6	May 28, 2019
2 Pallas	4.6 years	+6.5	34.8	April 6, 2019
3 Juno	4.4 years	+7.5	13.0	April 2, 2020
4 Vesta	3.6 years	+5.2	7.1	November 12, 2019
7 Iris	3.7 years	+6.7	5.5	April 5, 2019
433 Eros	1.8 years	+6.8	10.8	June 18, 2021
6 Hebe	3.8 years	+7.5	14.8	April 4, 2020
18 Melpomene	3.5 years	+7.5	10.1	July 1, 2019
15 Eunomia	4.3 years	+7.9	11.7	August 13, 2019
8 Flora	3.3 years	+7.9	5.9	May 11, 2019

Catching a close-passing asteroid usually requires a fairly large telescope—they rarely break +10th magnitude—and careful planning. **Parallax**, or the position of the asteroid against the starry background versus the position of the observer on Earth, comes into play. Also, most desktop planetarium programs fail to take deflection of the asteroid's path due to gravity into account. Your best bet? Use the JPL-Solar System Dynamics HORIZONS web interface to set your location and time, and then generate an ephemeris with right ascension and declination for the given object. Plotted out on a star map, you can even stalk the near-Earth asteroid as it flies past a bright star. These will often show motion in *real time*, like a distant, slow-moving satellite. What a view you would have sitting on such a rock, as the big blue marble of the Earth swung past. These can and do occasionally enter the Earth's shadow. I'll never forget following the slow drift of 4179 Toutatis in 2004 as it passed just four lunar distances from the Earth.

Stick around until April 13, 2029 (yes, that's Friday the thirteenth, for all of you triskaidekaphobes out there), and you can watch 99942 Apophis pass just 19,400 miles (31,220 km) from the Earth over Europe, Africa, and Western Asia, shining at +3.5 magnitude.

Good sites to monitor for upcoming Near Earth Object (NEO) passes are listed below.

- The IAU Minor Planet Center: https://minorplanetcenter.net/iau/lists/PHAs.html
- Space Weather: http://www.spaceweather.com/
- The Center for Near-Earth Object Studies (CNEOS) close approach list: https://cneos.jpl.nasa.gov/ca/

Clouded out, or just happen to live in the wrong hemisphere? The astronomer Gianluca Masi and the Virtual Telescope Project 2.0 generally produce live webcasts featuring close asteroid passes: https://www.virtualtelescope.eu/webtv/

ACTIVITY: HOW TO OBSERVE AND REPORT METEOR SHOWERS

Looking to do some real science? Good news: YOU can contribute to the science and understanding of meteor showers just by stepping outside tonight and counting what you see. Even in the modern era, amateur observations are a main source (along with radar observations) for understanding meteor showers.

For years, I've used a simple digital voice recorder to count and report what I see. This enables me to continuously keep eyes on the sky, lest we miss one, while reporting another. These days, most smartphones come readily equipped with free voice recorder apps. Simply note your sky's limiting magnitude (the faintest star that you can see), the time, location, which shower an observed meteor belongs to, and how many you've seen over the span of at least an hour. Some observers like to note other specifics, such as cloud cover, the estimated magnitude and detailed direction for each meteor, and thoughts and musings on life and the Universe, but the number over time from your location and the shower of origin is enough for a valid scientific observation.

The International Meteor Organization (IMO) welcomes these simple observations from meteor watchers worldwide. The IMO also maintains a handy quicklook page for ongoing showers, compiled using observer reports worldwide. This is a good gauge for ongoing activity, and a preview of what you might see tonight.

Happen to see a bright fireball? The American Meteor Society (AMS) keeps a handy log of fireball reports worldwide, and you can submit a report on their page as well. This is our first stop when we start hearing reports of a bright fireball worldwide across Twitter.

And speaking of social media, beware of spoofs. Folks love to cut and paste fake images and short GIFs from older reentries/fireball events in an attempt to pass them off as legit. Traditional meteor wrongs include the reentries of Mir and Hayabusa, the Chelyabinsk event, and scenes from the movie *Armageddon*.

IMO website: https://www.imo.net/

AMS website: https://www.amsmeteors.org/

Have DSLR, will travel . . . our own humble imaging rig: a DSLR with a wide-angle lens on a lightweight fold-up tripod and an intervalometer.

SOLAR OBSERVING

How to safely observe our nearest star.

"I simply can't stop observing when thinking that one can never know when nature will show us something unusual."

—Hisako Koyama, twentieth-century amateur astronomer and solar observer

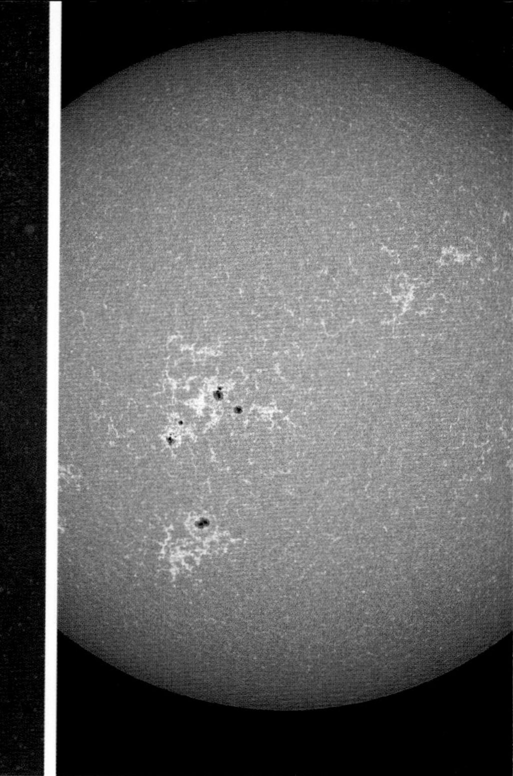

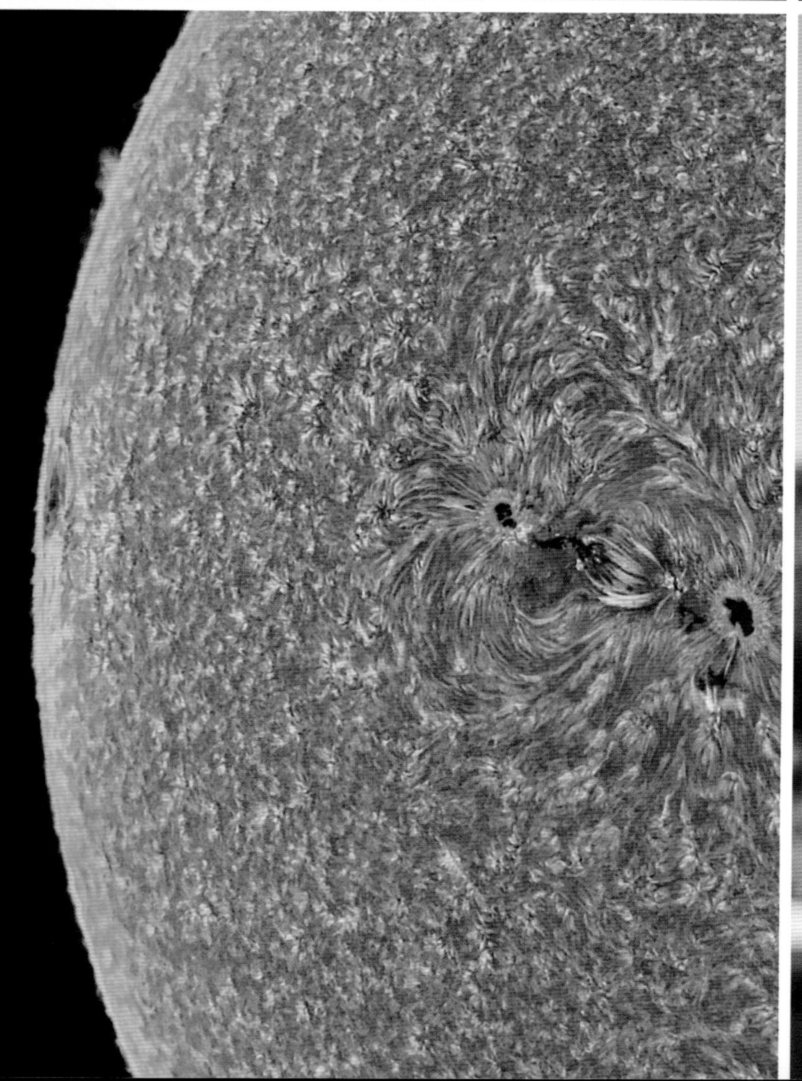

Astronomy doesn't end at dawn. With the right gear and filters, you can also observe the Sun. Unlike other faint fuzzies or tiny points of light, things are really *happening* on the Sun from day to day, sometimes from hour to hour. And unlike the Moon, the Sun rotates from our Earthly perspective, bringing new sunspots and activity into view.

ACTIVITIES

- **Daytime astronomical observing (page 147)**
- **Making a safe homemade solar filter for binoculars or a telescope (page 148)**
- **Making a Sun Gun solar projector (page 149)**

ASTRONOMY DOESN'T END AT DAWN.

GETTING TO KNOW OUR HOST STAR

Our Sun is a yellow dwarf G2 type star, about midway through its 10-billion-year main sequence life span.[1] The term "yellow dwarf" is a misnomer, though the Sun is indeed smaller than many of the stars out there in the night sky. For example, Antares would swallow up the inner Solar System all the way out to the orbit of Jupiter if you plopped it down in the center. However, red dwarf stars—the most common type of star in the Universe—only range from about 7.5 percent to 50 percent solar mass. It's a cosmic irony: Not a single red dwarf star is visible from the Earth with the naked eye.

No other astronomical body influences life on Earth like the Sun. The Sun shines via nuclear fusion, as the terrific pressure at its core fuses hydrogen nuclei into helium via the **proton-proton chain** nuclear process.[2] In a very real and intimate sense, we live inside the outer atmosphere of our Sun, as the **solar wind**, a steady stream of charged particles whips past the Earth, giving rise to **space weather**. Nearly every form of energy on the Earth beyond geothermal[3] can be traced back to the Sun; coal, for example, is the fossilized remains from plants and living creatures who ultimately gathered energy via photosynthesis when they were alive. The Sun also drives the wind and weather on the Earth.

Left corner: A massive sunspot on the Earthward face of the Sun, versus the puny Earth to scale (the blue dot to the upper left).

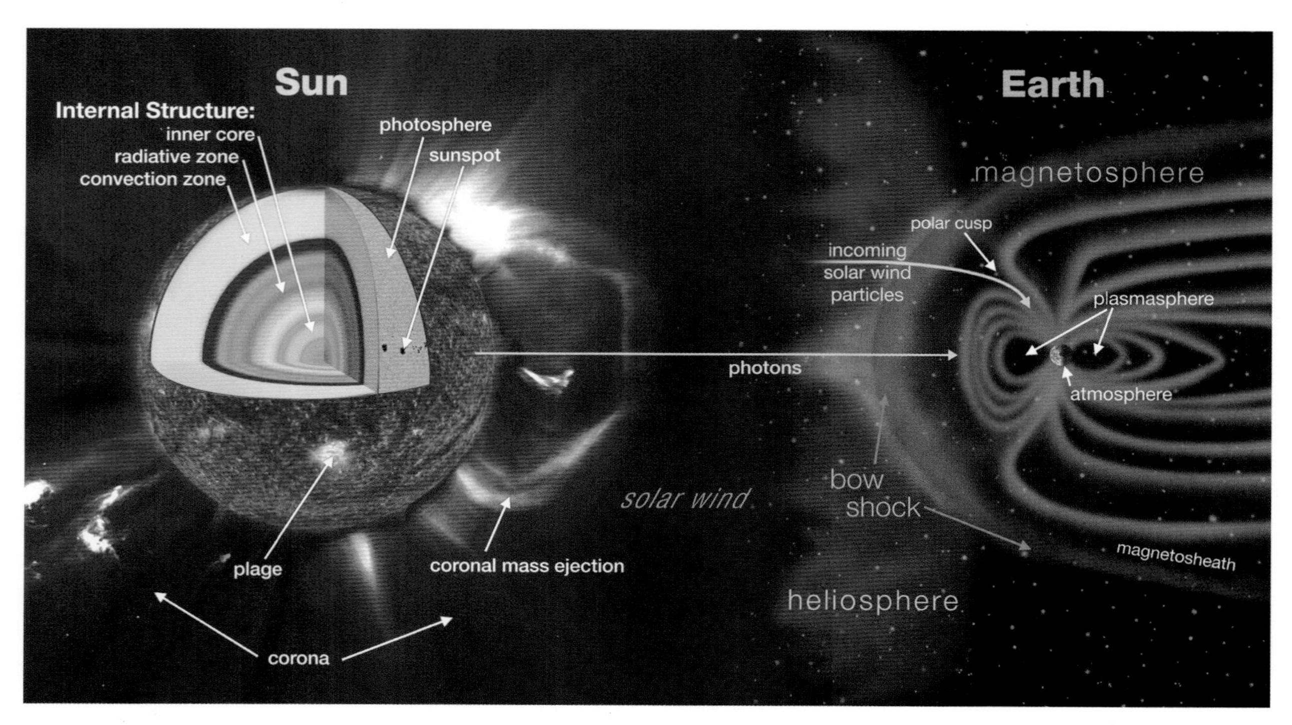

Living with a tempestuous host star: the interaction of the solar wind and space weather with the magnetosphere of the Earth.

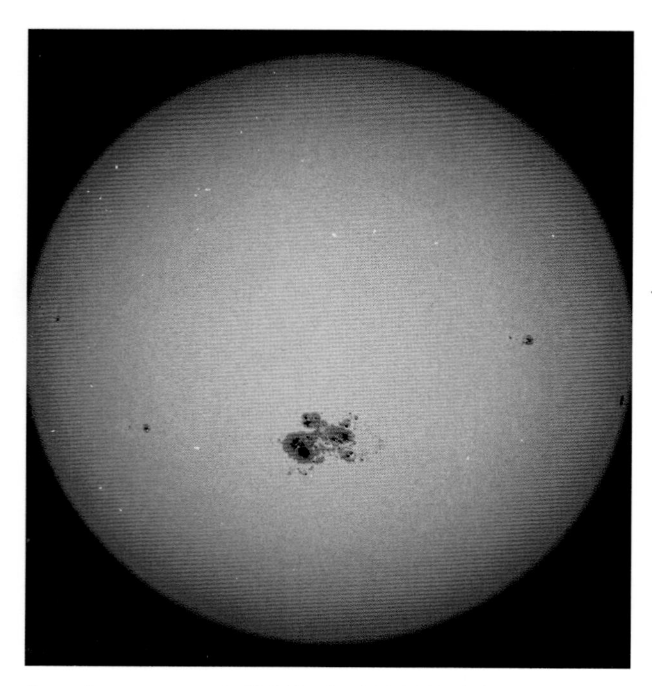

A massive sunspot nearly 80,000 miles (129,000 km) across graces the face of the Sun on October 23, 2014.

A BRIEF HISTORY OF SOLAR OBSERVING

Chinese astronomers kept records of large sunspots going all the way back to 800 BC.[4] Most of these were probably seen in pre-telescopic times while fog or clouds filtered out the Sun perched low near the horizon. The English astronomer Thomas Harriot was especially struck by the appearance of just such a massive sunspot in December 1610.[5] Harriot's famous contemporary Galileo projected the Sun onto a wall in a darkened room using a technique similar to a camera obscura,[6] and carefully sketched what he saw. Historic records of sunspots allow us to chronicle solar activity back over the centuries. For example, few sunspots were seen over the period known as the **Maunder Minimum**, which spanned from 1645 to 1715, a time of global cooling. Another shorter spotless period in the early nineteenth century included the infamous Year without a Summer of 1816. Japanese sketch artist Hisako Koyama made drawings of sunspots over more than four decades starting in 1944,[7] and solar astronomers at the Mount Wilson observatory have sketched the Sun every day since 1912.

You can even see the daily Mount Wilson observatory sunspot drawings at this website: http://obs.astro.ucla.edu/intro.html

Astronomers use a complex formula to derive a **daily sunspot number** from observed sunspot activity, a number that expresses current solar activity. More sunspots generally mean a more active Sun overall.

Why observe the Sun? Tempestuous space weather often impacts our modern technological, space-faring civilization. Solar activity can cause radio blackouts, GPS outages, and even cause airline crews and passengers to receive a higher-than-normal radiation dose. This is especially critical for crew on the International Space Station to know beforehand, so they can shelter near the core of the space station. Monitoring solar activity will become even more crucial during a crewed trip to Mars in the coming decades. A modern day Carrington Event (see Chapter 6: Near-Sky Wonders, page 104) could prove disastrous, killing satellites, knocking out power grids, and frying electronics. In 2009, the National Academy of Sciences and NASA released a study highlighting the need for an upgraded, hardened power grid to tackle just such an event.[8]

NASA and other agencies monitor the Sun worldwide and around the clock, from both the Earth and space. The GONG (Global Oscillation Network Group) is a solar observing network spanning five continents. High-profile solar-watching spacecraft include the joint European Space Agency (ESA)/NASA Solar Heliospheric Observatory (SOHO), ESA's Proba-2, the Japanese Aerospace Exploration Agency's Hinode, and NASA's Solar Dynamics Observatory (SDO). Watch the SOHO and SDO sites online to see what the Sun is up to, across the spectrum.

WHY OBSERVE THE SUN?

SOLAR OBSERVING SAFETY

First, let's talk solar safety. Solar observing is one of the most dangerous aspects of amateur astronomy: Even a brief glimpse of the Sun through an unfiltered telescope or binoculars can blind you permanently. Still, proper precautions for solar observing are straightforward, and I wouldn't let the potential dangers of observing the Sun scare you off, as solar astronomy is really an amazing pursuit. We like to make a habit of popping our equipment out for a quick look at the Sun daily, just to see what's up.

"Is that safe?" is usually the second question I get during public sidewalk solar observing, right behind, "What are you looking at?" with our telescope set up in the broad daylight, pointed toward the Sun.

The two basic methods for observing the Sun are via projection, or using a solar filter.

Projection simply means what it implies: aiming optics at the Sun and projecting an image onto a flat surface. This was the very first method early telescope users hit on, and it's a fun way to sketch sunspots, as you merely have to trace them out on a projected sheet of paper. Blank white sketchbook paper provides good contrast and works best.

Though this is the safest and simplest way to observe sunspots, there are a few caveats:

Viewing the Great American Eclipse of August 21, 2017, from the Pisgah Astronomical Research Institute (PARI) in the Smoky Mountains of Western North Carolina.

Projecting the Sun . . . during a transit of Mercury in 2016. Note: no image-stabilizing binoculars were harmed during this brief projection of the Sun.

- Make sure you're projecting the Sun in such a way that it's impossible for someone to duck down and look through the eyepiece, even for a split second . . . you would be amazed how many kids (and even adults!) will try to do this. I like to invert the right angle eyepiece holder downward toward the ground for projection, an awkward angle for anyone to attempt looking through it.

- Project the Sun far enough away from the eyepiece to render a decent-size image, but also be careful to not concentrate the light in a tiny spot . . . concentrated sunlight projected through a telescope can be intense enough to set paper on fire. Never leave a telescope projecting the Sun unattended . . . in fact, *never* leave a telescope unattended during public solar observing, period.

- Be careful to limit the time for a telescope projecting the Sun to no more than a minute, and be sure to give the telescope ample time to cool down in between. The heat from the Sun can scorch coatings, warp lenses and mirrors, melt plastic parts, and cause optical elements to come unglued. This is especially true of Schmidt-Cassegrain telescopes, where the sunlight passes through the optical tube three times. Better to use an old beater telescope for solar projection rather than a brand new, $10,000 rig.

- Adding a white-light solar filter to a telescope allows you to study sunspots at the eyepiece. Be sure to use a filter designed specifically for solar observing that fits snugly over the aperture of the telescope, to assure that the wind or prying hands cannot remove it. Visually inspect the filter before installation by holding it up to the sunlight: if anything shines through, it's unserviceable and needs to be disposed of and replaced.

Glass solar filters allow you to view the **photosphere**, the roiling surface of the Sun where sunspots form. Lunt and Meade/Coronado build small specialty telescopes made specifically for observing the Sun's outer **chromosphere** at a very specific wavelength of 656.28 nanometers known as **hydrogen alpha**, a region which displays streaming **prominences**, some looking like giant red acacia trees on the solar limb.

Incidentally, avoid using solar filters that screw on to an eyepiece lens. They're extremely dangerous, as they can overheat and crack without warning. These were actually standard issue on department-store telescopes right up until the early 1980s. We still occasionally see these old screw-on eyepiece solar filters at yard sales.

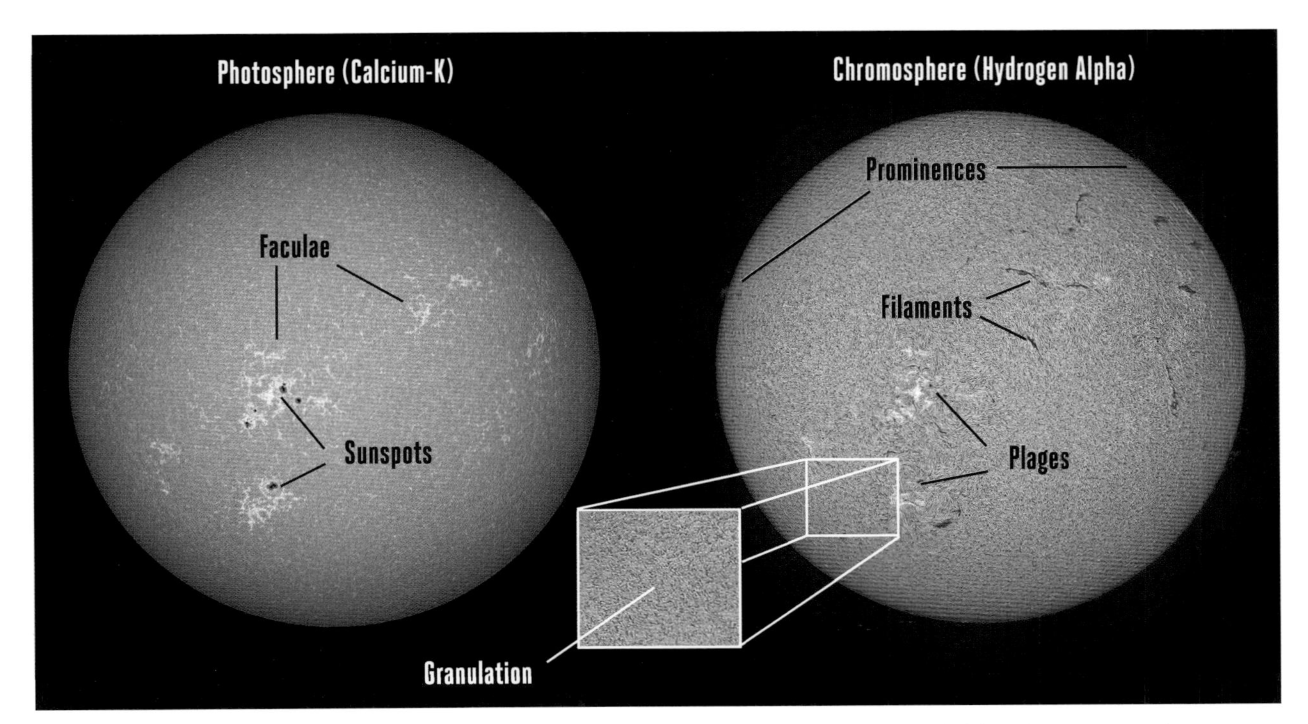

Two views of the Sun, showing the photosphere in Calcium-K (left) and the chromosphere in hydrogen-alpha (right). Both images were taken simultaneously to show the contrast between different types of features.

Other dangerous methods of solar observing from yesteryear include smoked glass, stacking exposed strips of film, stacking sunglasses, and oil-filled lenses. I once examined a set of nineteenth-century solar lenses at the Robert B. Ariail Collection of Historical Astronomy at the South Carolina State Science Museum meant for use with filtering oil; they were all cracked, every last one. Don't forget it isn't just visible light getting through those unorthodox filters, but a cross spectrum of dangerous invisible light, including the ultraviolet.

An impeding solar eclipse also heightens public interest in solar observing (see Chapter 3: Following Planets and the Moon in the Sky, page 46). Again, proper solar eclipse glasses carry the ISO 12312-2 rating.[9] Beware of counterfeits. Also, keep those solar eclipse glasses handy, as a large sunspot occasionally appears on the Earthward face of the Sun that's often visible without magnification.

But again, do not use solar glasses to look through any kind of optics at the Sun . . . they can't stand the heat!

Solar observing also presents other, more prosaic dangers that are often overlooked: sunburn and heatstroke. I had a persistent heat-induced headache after showing the public the November 8, 2006, transit of Mercury for several hours. Hydrate, use sunblock, and position your telescope near shade that you can duck under for a few minutes or so every hour and cool off.

THE STRUCTURE OF THE SUN

The dazzling photosphere of the Sun is what you see shining down on you on a hot summer's day. Below this realm of the sunspots are the **radiative** and **convective zones**, surrounding the denser core. The Sun has no proper solid surface, per se; it's just a ball of incandescent gas. Things cool down a bit in the **chromosphere** above the surface of the **photosphere**, though strangely, temperatures spike again to a few million degrees Celsius in the **solar corona**, glimpsed only briefly during a total solar eclipse. Researchers are unsure why, though powerful **nanoflares** are thought to blame for this extreme heating.

What to watch for: The best time of day to view the Sun is near **solar noon**, as the Sun transits the meridian highest in the sky through the least amount of atmospheric air mass, though the Sun always creates its very own atmospheric vortex due to heat. And hey, solar observing is the only time that light pollution is not a problem!

Aim your telescope at the Sun using its shadow. When the tube is casting its smallest shadow profile, you should very nearly have the Sun centered in the (filtered) view. Also, be sure to either cover or remove any finder scopes while solar observing. If not, you may get a nasty burn on your scalp reminding you to do so while you're bent over at the eyepiece!

Sunspots always appear in pairs with an opposing magnetic polarity. Like shadows, these scab-like growths creeping along the surface of the Sun have dark, umbral cores, surrounded by jagged penumbral edges. The largest recorded sunspot appeared on the Sun in March/April 1947. This monster-size sunspot was an amazing 187,000 miles (301,000 km) across, over twenty-three times the diameter of the Earth.[10]

Also, be sure to check out the peppery-looking texture of the photosphere overall, a phenomenon known as **granulation**. Brighter clear patches relatively free of **granules** in the photosphere are known as **facula** (*plural faculae*, Latin for "little torch"). Clear regions in the solar chromosphere related to faculae in the lower photosphere below are known as **solar plages**.

A solar telescope designed to specifically show off the Sun at **Calcium-K** wavelengths will show faculae and granulation in greater detail.

FUN FACT:

the Sun also flips magnetic polarity during every 11-year cycle.

HOW THE SUN CHANGES AND MYSTERIES OF THE SUN

As a large ball of gas, the Sun does not rotate as one solid sphere, but instead, rotates once every 25 days near the equator and once every 34 days near the poles. The Sun's rotational pole is also tipped 7.25 degrees relative to the ecliptic, and seems to nod its south pole toward us in March and its north pole toward us in September, though it's actually the Earth doing the moving as it orbits around the Sun. The Sun also appears 32 arcminutes, 32 arcseconds across in early January when the Earth is near perihelion, and 31 arcminutes, 28 arcseconds across when the Earth reaches aphelion in July.

Solar activity follows an 11-year cycle of maxima for sunspot activity versus minima. The hallmark of the start of a new solar cycle is the appearance of sunspots at mid-latitudes, which move toward the equator of the Sun as the cycle progresses. This pattern follows what's known as **Spörer's Law**. Richard Carrington first devised this rule, which was refined by the astronomer Gustav Spörer in 1861.

Fun fact: the Sun also flips magnetic polarity during every 11-year cycle. The proper full **Hale Cycle** for the Sun is actually 22 (11 times 2) years, as the same respective hemisphere returns to the same respective north-south polarity.

Why do sunspots never appear toward the poles? Why is the Sun's cycle apparently baked in at 11 years? Has it always maintained such a period, and will it always continue to be so? Are there longer cycles of solar activity? How normal is our host star, and how essential is its relative stability to the story of life on Earth? Clearly, there's lots we still don't know about the Sun, a continuing and fascinating astronomical subject of study.

ACTIVITY: DAYTIME ASTRONOMICAL OBSERVING

The Sun and the Moon aren't the only objects of study post-sunrise. I spent the last half of the 1990s stationed at Eielson Air Force Base outside of Fairbanks, Alaska, living in a town named (for real) North Pole. Living just below the Arctic Circle, astronomy was very much a seasonal spring and fall sport, a transition time when the length of the day was a somewhat normal 12 hours in length and the temperatures had yet to plunge to the sub-zero depths of winter.

The challenge in the summertime became on what date (usually in late July) I could first see a star or planet. This was usually a good sign that I could break out the telescope for the season.

But did you know that with a little practice, you can find planets and even brighter stars during the daytime? This feat of visual athletics requires knowing when and where to look; deep-blue, high-contrast skies; and patience.

Sometimes, you can simply start tracking a planet or star before dawn, and keep it centered in an eyepiece field of view using the telescope's right ascension clock drive post sunrise. Do not attempt to spot objects when they're closer than 20 degrees elongation from the Sun, as you run the risk of catching the dazzling Sun in the view. Amateur astronomers have managed to track Venus straight through inferior conjunction, and a bright comet could always make a visible appearance close to the Sun . . . but to make such an observation requires careful planning and safety precautions. To make such an observation, assure that the following conditions are met.

It's easy to see, once you know exactly where to look for it: Venus near the daytime crescent Moon (upper left).

- The Sun is physically blocked out of view, so that accidentally aiming the telescope at the Sun is impossible . . . and the Sun won't rise from behind the obstruction and suddenly come into view.
- The telescope's pointing accuracy is properly calibrated beforehand and dead on. In other words, there's no sweeping back and forth to find the target.
- You're using an extended hood on the aperture of the telescope to further block out any stray light.
- You use a camera at the eyepiece to project the view remotely onto a laptop screen.

Again, such an observation should only be attempted by a skilled observer, and is *only* feasible when observing bright objects such as Mercury, Venus, or a bright daytime comet (they do very occasionally occur) not closer than around 10 degrees from the Sun in the daytime.

Venus is by far the easiest object to spy during the daytime. The very best time to try is when it's near the greatest elongation and farthest from the Sun. A good time to nab any planet in the daytime is when the Moon is nearby as a handy guide. Strangely, the Moon has a much lower **albedo**, or reflectivity, than Venus, at 12 percent versus 70 percent, but it's also visually much larger, making it relatively easy to spot against the blue daytime sky. It's surprisingly hard to spy glittering Venus against the empty blue sky . . . until you see it. I've often started a star party a bit early with the Moon and Venus before sunset, and heck, Venus actually *looks* better at the eyepiece against a daytime blue versus nighttime black sky.

Jupiter is also a fairly easy grab before sunset. I once managed to see Jupiter 30 minutes prior to sunset while assigned to al Jaber Air Base, Kuwait. I simply noted the planet's position on the horizon versus a flag pole after sunset one night, then stood in the same place the next evening before sunset and started watching. I also managed to spy Sirius before dusk around the month of May when it's well-placed high in the southwest using the same method.

The same holds true for Mars near opposition: If it's near the gibbous Moon, give it a try. Likewise for any daytime occultations of the big four bright stars along the Moon's path (Antares, Spica, Aldebaran, and Regulus). These are all possible to nab near the daytime Moon using binoculars or a telescope. Even Mercury at greatest elongation should be possible . . . though sky conditions would need to be near perfect in terms of seeing and transparency.

Be sure to expand your astronomical repertoire into daytime observing . . . who wants to wait for sundown to begin an evening's viewing, anyhow?

ACTIVITY: MAKING A SAFE HOMEMADE SOLAR FILTER FOR BINOCULARS OR A TELESCOPE

You can spend over $100 for a high-quality, full-aperture, white-light solar filter for your telescope . . . or you could build one for less than $10. We've done so for DSLR lenses, binoculars, handheld viewers, and telescopes, large and small.

Supplies and Tools Needed

- Cardboard. Pizza-box cardboard works great for large scopes; tea boxes work great for binoculars
- Either a sheet of Baader AstroSolar Safety Film or (for small apertures like binoculars or a DSLR lens) an extra pair of approved solar safety glasses
- A box cutter or an X-ACTO knife
- A ruler
- Duct tape
- Super glue or a sturdy stapler

A homemade Baader safety film solar filter in use during a partial solar eclipse.

One sheet of Baader AstroSolar Safety Film can easily make multiple filters . . . and for a large telescope like an 8-inch (20-cm) SCT, the film doesn't need to cover the entire aperture (though the mask does). Simply cut a 60 mm offset in the two cardboard aperture covers, and stretch and glue the filter film sandwiched between the two. I wouldn't obsess over getting the film stretched too taut; most small wrinkles won't show up at the eyepiece.

Make sure the outermost cardboard mask is slightly larger than the aperture of the telescope. Now, you'll make a collar to join it up to the mask, using glue and duct tape. Cut 2- to 3-inch (5- to 8-cm)-wide strips of thin cardboard, then wrap them in concentric rings around the telescope to form fit it tight around the telescope tube. Once the mask is affixed to the collar, it should be snug enough that not only will no light get through, but the wind can't take it away.

Visually, Baader AstroSolar Safety Film gives the Sun a purplish overall cast, a noticeable but not objectionable tint. We've made these for nearly every instrument we've owned, and even a few for loaner scopes the evening before an event such as an eclipse or a transit of Venus or Mercury.

As with any solar filter that is commercial or homemade, be sure to visually inspect it before every use. Simply hold the filter up to the Sun and look for light leaks both through the filter and around the edges. Store the filter away in a dry place when it's not in use to avoid punctures . . . we noticed that the first filter we built in arid Arizona lasted for years, but the same style of filter would only last a season or so in damp Florida.

You can use the same essential concept for modifying a pair of solar eclipse glasses to fit on binoculars for safe solar observing. Again, these fit over the front aperture end, and you're sandwiching the crucial filter between two cardboard masks that attach to a mounting collar. I'm always on the lookout for a box or small container that comes ready-made for use in such a project.

You can watch my video of how I made a tea-box solar filter for our binoculars prior to the 2017 total solar eclipse here: https://youtu.be/cR9EgViRn4Y

ACTIVITY: MAKING A SUN GUN SOLAR PROJECTOR

Another possibility is constructing a Sun Gun or Solar Funnel projector using a funnel, some clamps, an extra metal-barreled eyepiece (not a plastic one that can melt!), and a projection screen made out of a thick but translucent (but not transparent) flexible material, such as a double-layer shower-curtain liner or rear-projection screen material. This will create a screen on the end of the funnel projected back from the telescope. The big plus from this is that it allows several people to view the Sun on the back of the projection screen at once.

The funnel will need to be cut to fit the eyepiece snug, with the smaller metal clamp drawn tight around it and perhaps shimmed up a bit with—you guessed it—duct tape. This sort of system works best on a simple small refractor with the funnel aimed straight back or through a right-angle eyepiece holder, though you might be able to get away with such a rig on a Newtonian for quick 1-minute views (remember the heating factor for catadioptric telescopes). Stopping down the aperture of a Newtonian reflector also helps combat overheating; you don't need all that aperture when you're viewing a large target like the Sun!

The key is in the focal length. The longer the funnel, the larger the solar disk will appear on the projection screen, though if it's too long, the image will also appear very dim . . . and if it's too short, the disk will not only appear tiny, but all that concentrated sunlight could puncture the screen! A plastic oil funnel around 18 inches (46 cm) long works best; the eyepiece selected versus the focal length for the telescope will determine the final image projection size, ideally around 100 mm or so. This is expressed as the focal length of the telescope (T) divided by 43 equals the focal length of the optimal eyepiece (e):

$T/43=e$.

Here are some suggested eyepiece focal lengths versus the focal length of the telescope used:

Telescope focal length = Eyepiece focal length

- 400 mm = 9 mm
- 600 mm = 14 mm
- 800 mm = 19 mm
- 1000 mm = 23 mm

Larger sun guns (sun canons?) have also been built using buckets, flower pots, and nearly anything else cylindrical you can imagine . . . but be wary of how much weight you're introducing at the eyepiece end of things. Amateur astronomers are a pretty resourceful bunch!

NASA has a great resource site on building a Sun Funnel: https://eclipse2017.nasa.gov/make-sun-funnel

A Sun funnel projector built out of a plastic soft drink cup.

ASTROPHOTOGRAPHY 101

Following the path from amateur astrophotographer to pro.

"Do, or do not. There is no try."

— *Yoda,* Star Wars: Episode V: The Empire Strikes Back

The debate is largely over.

The early twenty-first century was a turbulent time in the world of astrophotography. Purveyors of film still represented the pinnacle of the field, taking hours-long, carefully guided exposures with expensive gas-powered rigs. Into the fray came the upstarts. Some were armed with cookbook CCD cameras. Others with modified off-the-shelf webcams. Some simply aimed their smartphone cameras at their telescope's eyepiece and shot what they saw.

The intent of this chapter is to whet your appetite, and to introduce you to the diverse modern methods you can use to image what you see in the night sky. We're going to start with simple sky shots you can take tonight, either through a telescope or with a DSLR camera on a tripod, and take you up to entry-level deep sky astrophotography. This is the steepest part of the learning curve, going from shooting bright targets like the Moon or planets to faint deep sky objects. Jumping off into the deep end of astrophotography with its technical detail and jargon can be intimidating . . . but it doesn't have to be this way. If anything, technology has lowered the bar for entry into the exciting world of astrophotography, and you might just find that you have the gear kicking around to get started tonight.

ACTIVITIES

- Building a simple barn door tracker for night sky photography for under $10 (page 158)
- Smartphone astrophotography (page 161)
- Building a homemade planetary webcam (page 167)

A BRIEF HISTORY OF ASTROPHOTOGRAPHY

Just think, with DSLR in hand, you can quickly pass through all of the early stages of nineteenth-century astrophotography . . . in about one week. The Moon became the very first subject of astrophotography, when John William Draper successfully captured the first passable image of the First Quarter Moon on March 26, 1840.[1] Draper was working with a daguerreotype, exposing a tray of noxious silver iodide chemicals to the sky. The advent of collodion wet-plate photography meant astrophotography could go portable, and soon astronomers were using the newfangled technology to survey the northern and southern hemisphere sky. Those early efforts are archived online at the Harvard College Observatory Astronomical Plate Stacks (http://tdc-www.harvard.edu/plates/) with over half a million plates from both hemispheres taken from 1885 to 1993.

Why bother to save old glass plates in this modern age of astrophotography? Well, when astronomers noticed an anomalous brightening in Tabby's Star KIC 8462852 in the data from the Kepler Space Telescope, they looked back through archival glass plates from over a century ago[2] . . . but the mystery only deepened, as they found that Tabby's Star is actually *fading*. Here's another example: Smithsonian Center for Astrophysics researchers announced in 2016 that astronomers had the evidence to make the case using glass plate spectra for a potential exoplanet system around Van Maanen's Star way back in 1917 . . . if they'd only known to look for it.

Other astrophotography milestones fell in quick succession, including the first sunspot photo (1845), the first image of totality during a solar eclipse (1851), and the first capture of a meteor (an Andromedid) by the Austro-Hungarian astronomer Ladislaus Weinek (1885). Draper also managed to capture the first decent deep sky image of the Orion Nebula in 1880.[3]

The very first successful image of totality taken during a total solar eclipse. This daguerreotype capture was obtained by Johann Julius Friedrich Berkowski on July 28, 1851, from the Royal Observatory in Königsberg, Prussia.

We like to briefly mention the historical course of astrophotography because, again, it plots the learning curve of many a budding astrophotographer, from bright, easy targets, to fainter ones requiring more technical expertise. Plus, the marriage of photography with that other essential nineteenth-century innovation—the spectroscope—marked the true transition from observational astronomy to astrophysics, as astronomers could, for the very first time, directly sample the composition of the cosmos.

Okay, so lots of those early efforts at digital astrophotography were messy, and heck, they had all of the same flaws as early digital imaging. But nowhere has Moore's Law been more applicable than in digital photography. The rise of DSLR cameras has not only lowered the entry level bar for astrophotography, it has also virtually taken over the field. Plus, the widespread use of full frame sensors over the last decade truly marked the end of the film era, as sensor size finally reached the hallowed 35 mm size used for over a century in film cameras.

WHICH DSLR IS RIGHT FOR YOU?

Today, the burning question of "which DSLR should I purchase?" is right up there among those big life decisions for astrophotographers, along with choice of profession, where to live, and whom to marry.

What's more, astrophotography has diversified. You can now couple a DSLR directly to a telescope, piggyback it to take advantage of the telescope's tracking, or simply set it on a tripod to take wide-angle shots of the sky. You can modify a webcam to shoot planets, or use a low-light security camera to shoot video. You can even install a wide-angle, all-sky camera to continuously monitor the sky for meteors or auroras. Speaking of auroras, you will want to use short exposures on these, as they can change quickly and get amazingly bright enough to cast shadows.

HOW A MODERN CCD/CMOS WORKS

Amateur astronomers are usually only about a decade or so behind the professionals when it comes to technology. Take a look at some of the top-notch photos of the planets in astronomy textbooks from the 1960s and 1970s; today, a backyard astrophotographer would throw away (or simply delete) such grainy images without a second thought.

A digital imager works by converting incoming photons into electronic signals. Commercial CCD (charge-coupled device) cameras came on the market in the early 1990s. The rise of CMOS (complementary metallic oxide semiconductors) along with active pixel sensors for noise reduction (the primary drawback for early CMOS detectors versus CCD cameras) has really brought the price of high-end DSLRs down.

DSLR sensors work by light binning, or filling individual buckets with data from incoming photons. Some of the older DSLRs, for example, treated stars as artifacts in their noise reduction algorithms, and promptly erased them.

The differences a little post processing can make: Messier 42, the Orion Nebula, before and after.

KEY CAMERA SETTINGS

Modern DSLRs borrow much of their terminology and settings from the earlier era of film.

ISO

Here's a fine example of an acronym that now has nothing to do with what it actually stands for. ISO means International Standards Organization, which probably refers to its adoption as a standard measure of calibration in the early days of photography. Simply put, ISO refers to light sensitivity of exposed film, or these days, the sensitivity of the camera's imaging sensor. Low ISO is 100 to 200, good for an average bright daytime scene. In astrophotography, however, you're forced to use higher ISO settings up through 3200 or 6400 for dim objects. In the days of film, ISO 3200 or higher was difficult to achieve, as few vendors sold this sort of specialty film.

As with anything, however, there's a trade-off in terms of going deep . . . in the olden days of film with a preset ISO per roll, this also meant a grainier image at high ISO. Today, the digital situation analogous to film grain is noise. Noise reduction settings help with this somewhat, though a good rule is to not only use higher ISOs for darker settings, but also to use the highest setting you can get away with without introducing additional unwanted camera noise. If you're stacking images, you may want to turn off automatic noise reduction entirely.

F/RATIO

Another aspect of telescopes is familiar to photographers: focal ratio, sometimes referred to as "f/stop." This is simply the ratio of the aperture versus the travel distance from the primary collector to the eyepiece. For example, a 200 mm aperture mirror with a 2000 mm focal length (typical for an 8-inch [20-cm] SCT) has an f/stop of f/10. This matters because a low or fast focal length (say f/6 to f/4) will yield a generous field of view and fast light collecting potential—great for deep sky observing and astrophotography for faint extended deep sky objects—but isn't splendid when it comes to small bright objects like planets. A longer focal ratio cuts these bright point sources down, and easily yields higher magnifications.

One factor that's never changed from film to digital is how fast an optical system is. This is simply the ratio of the aperture versus the focal length of the camera, or the distance from the lens to the image sensor. A larger aperture and smaller f/ratio means shorter exposure times, better resolution, and a brighter overall image. Cameras change f/ratio either by stopping down the aperture or changing the focal length by use of a zoom lens. This is also why a 400 mm-wide aperture lens can cost ten times the price of a smaller aperture model with the same focal length. **Prime lenses**, lenses with fixed focal lengths, are preferred for astrophotography, as they deliver tack-sharp images across the field, free of aberration or field and color distortion. Another way to combat chromatic aberration is by using smaller apertures and longer exposures, which of course is a trade-off that requires precise tracking. Zoom lenses, while convenient for travel and multipurpose use, also frequently cause the focus to shift.

SHUTTER SPEED

This setting fixes the amount of time the shutter is open and the exposure length. For astrophotography, you'll often use the M (manual) or Bulb setting. The good news is electronic shutter systems aren't prone to the camera-jittering shutter slap inherent in older mechanical systems.

WHITE BALANCING

Our eyes evolved to see the light on a noon day scene as essentially white, though in reality, it's a mix of colors. Auto white balancing on modern DSLRs handles this situation under dark skies pretty well, and we wouldn't get too obsessed with alternate white balance settings unless you're shooting with a modified DSLR and going after, say, faint hydrogen alpha emission nebulae. Plus, white light and color balance can always be corrected in post processing using Photoshop or your software of choice, if you shoot in RAW format.

Many astrophotographers also prefer shooting in RAW, versus a compressed file format such as JPEG. This ensures no image fidelity is lost due to compression, but it also makes for a huge file size. Keep in mind that RAW is also not a universal format compatible with all image processing software.

One neat beginner trick is to use the Tungsten white balance (2,800 Kelvin) setting for wide-field star photos to counter the yellowish tint of light pollution.

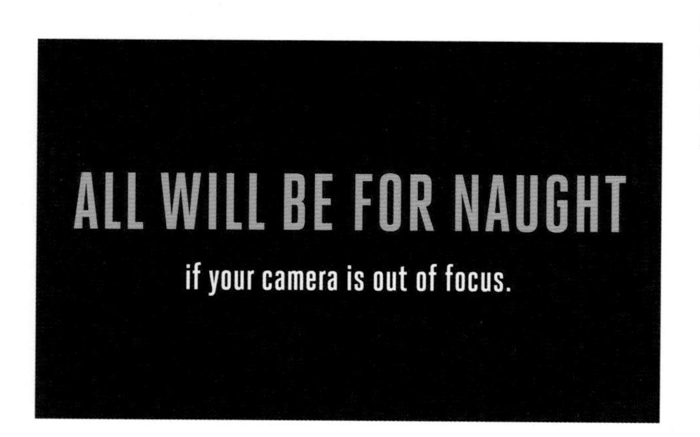

FOCUSING AND FRAMING

All will be for naught if your camera is out of focus. Focus is actually tough to nail down . . . seeing and the burbling of the atmosphere above often foils the effort when it comes to achieving a fine focus.

Shooting through a telescope, you'll be using the instrument's focus; piggybacking or shooting on a tripod, you'll use the manual focus on the camera's lens. This is where live preview comes in especially handy. Set the ISO high, the aperture wide open, and the exposure to Bulb or Manual setting. Then, focus the stars in the field down as sharp as possible. If no bright stars are available, you may have to aim at one, focus, lock the focus down, then recenter on the dimmer target and shoot a few test shots to assure acquisition. The Moon or a bright planet can also make a great target to fine focus on. Remember to recalibrate this occasionally, as focus can shift slightly as the camera and the telescope cools down over the course of a night.

The first difficulty many starting astrophotographers encounter is fine focusing under dark lighting. Again, live preview is essential for focusing, though this feature is now pretty much standard on most DSLRs. Live preview also helps during alignment and focusing, and don't be afraid to shoot a few throwaway shots to get the framing right before getting down to business. A good way to frame up a faint target after focusing is to use a guide scope.

Here's one nifty trick for getting a quick fine focus: Switch to a super high ISO setting, then take a short exposure just to check and see if you need to tweak the focus. Once you're dialed in, all you need to do is flip back to the proper ISO setting and exposure times.

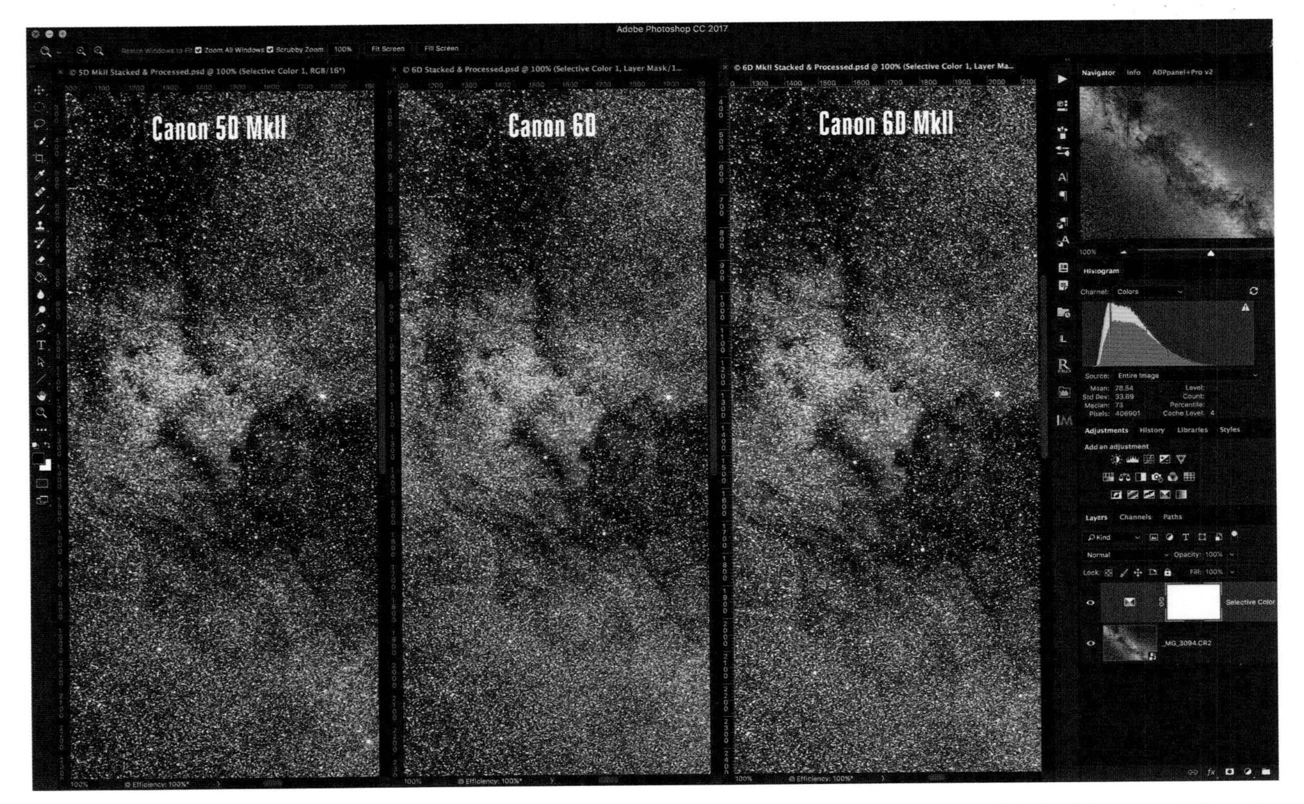

Why modify a DSLR camera? This sequence of three images demonstrates the benefits of modification to bring out subtle detail in an extended target such as the North American Nebula. The final combined images show a modified DSLR (far left) versus two unmodified, off-the-shelf DSLRs. The modified DSLR is sensitive to deep reds emitted by hydrogen gas.

TO MODIFY, OR NOT TO MODIFY?

One key problem when using a DSLR for deep sky photography is the infrared cutoff filter. A stock camera with the standard factory filter gives nebulae a pale, dim cast, taking out some of those splashy reds we all love. Many deep sky astrophotographers get around this by modifying DSLRs, swapping out the standard wide-band filter for a narrow band one that keeps the infrared out while letting those deep reds in. Some companies, such as AstroHutech (http://www.hutech.com/) offer modified, deep sky-ready DSLRs. We'd think twice about doing such a modification on a beloved $2,000 camera ourselves.

Keep in mind that a modified DSLR is also compromised for daytime use, something to consider if your DSLR does double duty as the family camera. You can create a custom white balance setting to offset the pinkish hue such a modification will create.

Canon actually offered two modified models, the 20Da and 60Da on limited production runs some years back; Nikon has also entered the fray with its D810A model camera, a full-frame DSLR with a specialized IR cut filter specifically for astrophotography.

IMAGE STACKING AND PROCESSING

Image stacking is an amazing technology that is becoming an increasingly automated process. A stacking program will take a series of images, align them, then filter through the very best ones, searching for moments of true fidelity where microseconds of steady seeing might bring out the very best details.

Stacking reduces the noise factor in an image exponentially; the more images in a stack, the less noise (unwanted variations across the image) and the more signal (actual detail you want) gets through.[4]

Early efforts would see astrophotographers painstakingly sift though sometimes thousands of frames, often after doing equally tedious manual alignments, again, frame by frame. Stacking programs will now sift through these, allowing the user to select, say, a cutoff of the best 10 percent or so of images. This makes planetary detail really pop out and means that what used to take several hours' worth of guiding for a deep sky shot can now be achieved by shooting a quick series of a few dozen 15-second exposure frames.

Registax is a great (and free!) industry standby for stacking and processing: https://www.astronomie.be/registax/

Another newcomer in the deep sky stacking game is the free program Deep Sky Stacker: http://deepskystacker.free.fr/english/index.html

TAKING ON TRACKING

If you're taking shots much longer than 30 seconds wide-field (e.g., shooting the Milky Way), you'll need to track the sky as the Earth rotates. To do so, you'll need precise **polar alignment**. Any image taken through a telescope other than a quick shot of the Moon will require tracking of some sort. The good news is, getting a precise alignment and tracking with modern GPS-controlled mount systems is easy, and **periodic error correction**—a correction against backlash, or the mechanical error of your given clock drive for the telescope's polar mount—can now often be corrected for via software.

Remember the handy 500 Rule for tracking: 500/focal length = time in seconds you can get away without tracking, before stars begin to trail.

If you're taking extremely long exposures (measured in hours), you will also need a secondary guidescope to help the exposure along.

One program that does this is PEMPro—a 30-day free trial is available from CCDWare at http://www.ccdware.com/products/pempro/features.cfm

REMEMBER THE HANDY 500 RULE FOR TRACKING:

500/focal length = time in seconds you can get away without tracking, before stars begin to trail.

Desert oasis on a small planet: The Milky Way rising over a small lake on the outskirts of the Atacama Desert in Chile.

ACTIVITY: BUILDING A SIMPLE BARN DOOR TRACKER FOR NIGHT SKY PHOTOGRAPHY FOR UNDER $10

Looking to really bring in astrophotography under budget? It's possible to build a simple hinge or barn door tracker for $10 using parts from your local hardware store. This is a mechanical unit, meaning you'll simply be turning the screw one quarter turn every 15 seconds to track the sky.

This assumes you already have a tripod and a DSLR. One other key piece is the camera ball-head to mate the camera (camera adapter sockets are ¼-inch [6-mm] standard) with the hinge. I picked one up from Amazon for the princely sum of $5.

Supplies and Tools Needed

- 8-inch (20-cm) strap hinge
- ¼-inch (6-mm) (20 turn ½-inch [13-mm]) screw (to attach the ball-head to the hinge)
- Large washer (glue this to the underside of the hinge, for the drive screw to push against)
- 2-inch (5-cm) long screws (10-32)
- Nuts and washers (10-32)
- ¼-inch (6-mm) Acorn nut
- One large fender washer
- One wing nut
- Super glue

A simple barn door tracker, affixed to a camera tripod along with a DSLR camera, ready to go for a night's worth of wide-angle sky imaging.

Feel free to vary on the design . . . for example, many trackers call for a large cardboard or plastic wheel for turning affixed to the main screw, but we found that a simple wingnut works fine. Just be sure to turn it gently so as to not jingle the camera during the exposure.

Two musts: Make sure the hinge has the camera mount hole located about 7 3⁄16 inches (18.3 cm) from the hinge joint, and the turn screw used is threaded for thirty-two turns per inch . . . otherwise, the ¼ turn per 15 seconds rotation rate will be off.

Assembly: The ball-head attaches to the upper hinge, and the lower hinge attaches to the tripod-head. The drive screw will simply push against the top flange of the hinge during use. Make sure you zero out the drive screw down to flush at the beginning of an exposure, to assure you don't reach the end of its travel length early on in the exposure.

Super glue the fender washer under the upper hinge, and use the acorn nut to attach the assembly to the tripod. Finally, install the wing nut on the turn screw.

Use: You'll want the hinge of the tracker pointed directly at the north rotational pole. I actually cheat a bit, and simply clip a green laser pointer along the axis of the hinge for a quick simple alignment in the field. A metronome app also works great for calling out 15-second intervals during an exposure, hands free. This sort of barn door tracker will function and track accurately for about 10 minutes before tangent error throws it out of whack.

And as low tech as it is, the NASA astronaut and avid amateur astrophotographer Don Pettit actually assembled a similar modified camera rig driven by a hand drill for use on the International Space Station![5]

DRIFT ALIGNMENT METHOD—
A QUICK RUNDOWN

Here's how to manually achieve polar alignment the old-fashioned way.

Using an equatorial mounted telescope, level the mount, then aim the tube straight back along the right ascension axis toward Polaris in the northern hemisphere,[6] or +5.5 magnitude star Sigma Octantis in the southern hemisphere. Some German equatorial mounts actually come equipped with a polar alignment scope built through the right ascension axis, complete with an illuminated reticle etched with the layout of the northern celestial pole versus Polaris and Ursa Major, a handy feature.

This sort of rough alignment is enough for casual observing or short (less than 3-minute) astrophotography; in fact, I wouldn't get obsessed with getting a more precise alignment at, say, a public star party, as lines to look through the eyepiece build up around the block.

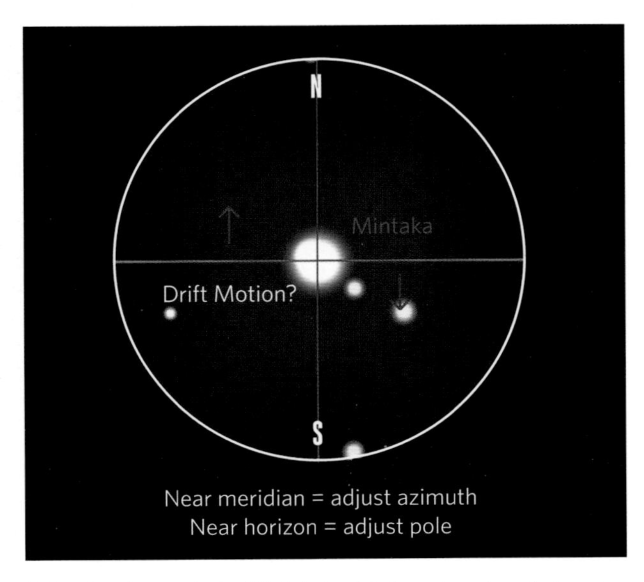

Drift polar alignment, simplified. A simulated eyepiece view of a target star near the celestial equator in the Belt of Orion. Drift observed to the north or to the south when the star is near the meridian means the azimuth of the mount needs to be adjusted, while drift when the star is near the horizon means the elevation of the mount's pole needs adjustment.

For longer exposures, you'll need to use the **drift alignment method**. Many first-time neophyte astrophotographers are intimidated by this seemingly complex procedure, which is really quite simple:

1. Roughly polar align the telescope.
2. Aim the telescope at a bright star within less than 20 degrees declination of the celestial equator near the meridian.
3. Using an illuminated reticle eyepiece, center the star and nod the scope up and down in declination. If you don't have an illuminated eyepiece, you could simply move the star to the very edge of the northern or southern field of view.
4. Align (rotate) the eyepiece so the reticle crosshairs are aligned, north-south, east-west.
5. Center the star again and watch as the tripod's motor tracks the sky. If the target star drifts **north**, adjust the telescope's azimuth on the mount **east**. If it drifts **south**, adjust it **west**. It may take a few fine adjustments to remove the drift entirely.
6. Now, we need to level the polar axis. Aim at a star near the eastern horizon also 20 degrees declination from the celestial equator, and watch. This time, if the star drifts **north**, **lower** the polar axis a tiny amount. If it drifts **south**, the polar axis needs to be **raised** a tiny amount.

(See an in-depth look at polar alignment method at: http://astropixels.com/main/polaralignment.html)

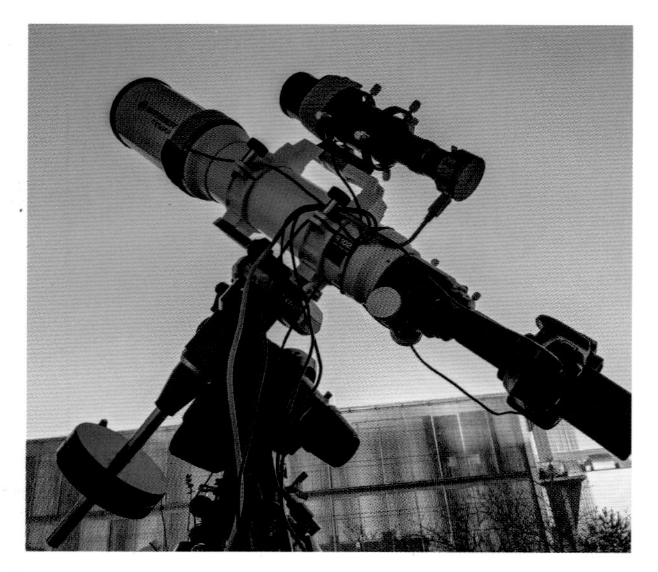

An example of a star-tracking guiding rig, piggybacked on an equatorial mounted refracting telescope. Note that the guidescope is connected to the bracket with duct tape!

A QUICK GUIDE TO GUIDING

Even after you get a precise polar alignment, you'll need to guide the telescope along for exposures much longer than a minute, or stars will smear out in an unwanted trail over the image. Motor backlash, gear slippage, shifting weight as the telescope moves, and mount flexing due to temperature changes can all throw a carefully aligned telescope slightly off-kilter.

This is where using a dual imaging rig or auto-guiding with an off-axis guider can come in handy. You're either viewing a guide star through a secondary piggyback-mounted telescope, or using a CCD camera and an autoguider camera to keep a target star centered, freeing you up to do other things (or catch some quick sleep!). PHDGuider is a great resource for autoguiding: http://www.stark-labs.com/phdguiding.html

DEDICATED ASTRONOMICAL CAMERAS

Interested in capturing meteors or time-lapses of the sky throughout the night? Many amateur astronomers now have dedicated all-sky cameras mounted in their backyards.

Whenever a video of a bright meteor flashes across social media, it's always from a video source that was up, ready and running at the time, such as a car dashcam, a security camera, or a mobile phone recording another event. All-sky cameras running through the night can also serve as unwavering sentinels to capture the late-night sky scene.

An all-sky camera incorporates a camera, a fisheye wide-angle lens aimed at the zenith, and a clear plastic or glass dome for a weather enclosure.[7]

VIDEO ASTROPHOTOGRAPHY

Low-light security cameras can also be repurposed for video astrophotography. These are especially handy for catching asteroid occultations of stars. However, a standard DSLR running video works just fine to document occultations of planets or bright stars by the Moon. Just make sure to balance the exposure between the bright Moon and the star or planet, so you have detail on the Moon, without losing the star or planet in its glare. This can be tricky if the Moon is near Full phase, and you might just have to settle with letting the glare of the Moon blow out the image just a bit.

A portrait of the modern solar system . . . all captured with a smartphone coupled to a telescope. From left to right: The Sun in hydrogen alpha, Mercury, Venus, Mars near opposition, Jupiter (note the double moon shadow transit), Saturn with its rings tipped wide open, and tiny Uranus and Neptune (sorry, Pluto!).

ACTIVITY: SMARTPHONE ASTROPHOTOGRAPHY

Don't own a DSLR? Smartphone cameras have increased in quality to the point that amateurs are now getting pretty decent images of the Sun, Moon, and brighter planets, simply by aiming their phone camera at a telescope's eyepiece.

You can also image bright constellations, ISS passes, and Iridium flares using a smartphone mounted on a tripod clamp, along with night sky capture software such as NightCap Pro for iPhone.

"My astrophotography began when I found myself with an iPhone and a Twitter account," says astrophotographer Chris Becke. "I saw people also putting their iPhones up to their telescope [with acceptable results]."

To this end, there are now ready-made clamps on the market that will enable the user to mount their phone over a telescope eyepiece. As with DSLRs, the key to smartphone astrophotography is manual camera control for exposure time, brightness, and contrast. With most phones, you can simply hold your finger on the image and it will then open up additional exposure controls. Medium magnification works best for focusing and acquisition, as you're basically afocally shooting through the telescope and trying to align the exit pupil of the eyepiece with the entrance pupil of the camera.

An iPhone coupled to the eyepiece of an 8-inch (20-cm) Schmidt-Cassegrain telescope equipped with a solar filter during a solar eclipse.

LANDSCAPE ASTROPHOTOGRAPHY

An easy entry-level way to start into astrophotography is to shun the telescope entirely and shoot landscape shots along with the night sky on moonless nights. This is the next logical step after shooting star trails, and some top astrophotographers find a niche doing nothing more than shooting the plane of the Milky Way galaxy over iconic natural sites.

A simple barn door tracker (see Building a Simple Barn Door Tracker, page 158, earlier in this chapter) is a great way to capture such a time lapse; you can add in the exposed foreground on a final, untracked shot. Flash a red light across the foreground briefly to give it an ethereal, otherworldly glow, or you can simply compose a shot with the silhouette of a lone person with a telescope, contemplating the Universe.

Do make an effort to journey to as dark a site as possible, as avoiding light pollution and unwanted sky glow is definitely worth the effort. Weekend camping trips near New Moon are a monthly ritual for many an astrophotographer.

Modern computerized mounts have sparked a revolution in modern wide-field astrophotography. These battery-powered trackers mount between the tripod and the DSLR for dedicated sky tracking. Precise polar alignment is even possible using the polar alignment scope and a smartphone app. These mounts are a great upgrade from a homemade barn door tracker, and only cost about $250, about what you would spend to build a motorized barn door tracker from parts.

...ZODIACAL LIGHT—

the ghostly glow of illuminated dust along the plane of the ecliptic.

MILKY WAY PHOTOGRAPHY

Milky Way Season: The core of our Milky Way Galaxy sits highest due south (from the northern hemisphere, transiting due north from the southern) at dusk right around the September equinox. The months prior to this are a good time to catch the Milky Way later in the evening, before it gets swapped out behind the glare of the Sun.

One app named PhotoPills used for Milky Way shooting (http://photographingspace.com/photopills-for-milkyway/) will even overlay the alignment of the plane of the Milky Way with the foreground, allowing you to plan out your imaging session in advance.

I wouldn't pack that camera away from October to March, however, as you still have a good view up north of the outer Perseus Arm and Orion Spur extension of the Milky Way visible during winter months, an interesting view in its own right.

Use as fast a lens as possible for Milky Way astrophotography, f/2.8 or even f/1.4 if you can. Shoot as long an exposure as you can get away with without stars trailing. For a wide-angle 14 mm lens, you can go up to 30 seconds, then 10 to 15 seconds for a 35 mm. Use high ISO settings all the way up to 1600, 3200, or even 6400 if needed. Try doing that with film!

Another phenomenon can be captured along with the band of the Milky Way near either equinox, as the **zodiacal light**—the ghostly glow of illuminated dust along the plane of the ecliptic—creeps up from the sunward horizon in the predawn or post-dusk sky. In the northern hemisphere, the angle of the ecliptic favors sighting the zodiacal light near dusk in March and dawn in September; the reverse is true for observers in the southern hemisphere. If your skies are really dark and moonless, you might also want to try and capture the **gegenschein**—the phantasmal counter-glow of sunlight glinting off of dust focused at the anti-sunward point—high in the sky near midnight.

As the Golden Hour gives way to the Blue Hour: the Milky Way emerges over Crater Lake National Park in Oregon.

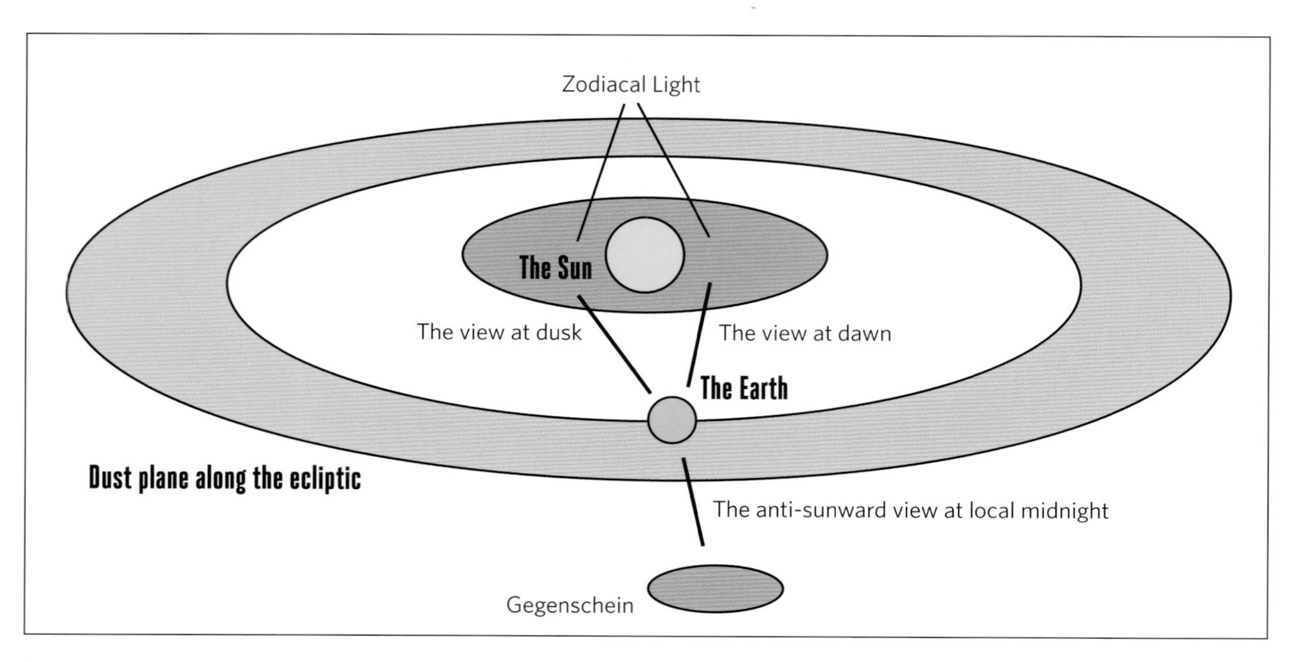

Dust scattering light along the plane of our solar system: the view from Earth along the ecliptic plane looking sunward toward the zodiacal light and the elusive counterglow known as the gegenschein at the anti-sunward point (okay, I've never seen it, either . . . you need pristine skies free of light pollution).

ON TRAVEL ASTROPHOTOGRAPHY

(AND HAVING AN ULTRA-PORTABLE RIG)

As with observational astronomy, travel versus astrophotography presents its trade-offs. If you're on an extended sabbatical and backpacking, for example, you want to follow the mantra of "pack light, pack right," though you won't want to miss out on any astronomical opportunities. I carry a DSLR, an ultra-portable but sturdy Dolica Proline tripod, a set of Canon IS 15x45 binoculars, a single 18–200 mm zoom lens (breaking the fixed lens rule, I know, but it allows us to only use and carry one lens), and lots of SD cards.

Our general purpose, lightweight travel rig. Pictured: binoculars, DSLR camera, collapsible tripod, intervalometer, zoom lenses batteries, SD cards, cables, tools, red/white headlamp, and oh, yes, bug spray and sunblock!

Other good hardware to have: a T-adapter for telescope mounting, a right-angle viewer (to eliminate the literal pain in the neck of having to continually contort under the camera and look through the viewfinder), an intervalometer for taking series of timed shots, and a wireless remote to perform the duty of a remote cable release.

Photographer's Ephemeris (http://photoephemeris.com/) is another great place to plan shots, especially if you're trying to catch the moonset/moonrise behind a foreground object. The Moon and the Sun are actually fairly tiny subjects to photograph, only half an angular degree across. You see this on social media all the time when folks shoot the Full Moon with their smartphone alone, resulting in a tiny white dot. If you're trying to catch the Moon rising behind a foreground building such as a castle, for example, you'll have to be positioned on the order of 3,281 feet (1,000 m) away or more to make both appear roughly the same size.

Shooting **star trails** is a great place to start into panoramic astrophotography. Simply wait until a moonless night, aim a tripod-mounted DSLR with a wide field of view at the north celestial pole, and open the shutter for about an hour or so. Or better yet, take consecutive 30- to 45-second exposure shots (to avoid getting swamped out by sky glow and light pollution), then stack them together in post processing. This will capture concentric rings and arcs of star trails made due to the rotational motion of the Earth. Shooting perpendicular to the rotational pole can be an interesting exercise as well, as you capture the rising and setting paths of the stars and planets. You want to use higher ISO settings for larger f/stops and consecutive 30-second exposures. For example, an ISO of 400 would work at f/1.4, then I would double the ISO for every whole f/stop number afterward for star trails, so as to not over saturate the image. Unlike film, a digital sensor (almost) never forgets a photon.

Generally speaking, you want to try and use the lowest ISO you can get away with. If you're shooting a continuous exposure and not a consecutive series of 30-second frames, start at short 3-minute exposures, then double them until you get the right mixture of f/stop, ISO, and exposure time for the local sky conditions.

CAPTURING TOTALITY: SOLAR ECLIPSE ASTROPHOTOGRAPHY

Ready for the next total solar eclipse crossing the United States on April 8, 2024? Solar eclipse photography during the long partial phases of an eclipse is pretty straightforward; you're simply shooting the Sun through a front-of-the-lens mounted, safe solar filter (see Chapter 8, Solar Observing Safety and Making a Safe Homemade Solar Filter, pages 143 and 148). During totality, however, filters can come off, allowing you to capture crucial features such as the pearly corona and vast prominences leaping off the limb of the Sun. Totality only lasts scant minutes in duration,[8] not the time to worry about fretting with camera settings . . . you want to be ready to start shooting, right out the gate.

Now, I'm going to get slightly mathy, if only for a minute. I could get mathier, trust me.

A 200 mm focal length is really the minimum size for a standard cropped sensor DSLR needed to produce a Sun that's more than a dot in the image, and about 1,000 mm focal length is desirable for a good-size composition. Multiply this focal length by about ⅔ (66 percent) for a full-frame DSLR.

A good exposure formula for totality is: *t=f^2/(Ix2^Q)*, where:

t is the exposure time;

f is the focal ratio;

I is the ISO;

and *Q* is the brightness exponent. (Baily's Beads=11, chromosphere=10, prominences=9, and the corona out to one solar radii=7)

For example, an f/8 focal ratio aiming to capture the corona (Q=7) at ISO 400 would work out to;

t=8^2/(400x2^7) which simplifies to:

t=64(400x128) or, calculator app out: 64/51,200 = 0.00125, or 1/0.00125 =1/800th of a second.

Totality! An amazing sequence featuring totality from the August 21, 2017, total solar eclipse, spanning the first contact "diamond ring" phase through Baily's Beads, the pearly white corona and prominences on the solar limb during mid-totality and last internal contact. The totality images are a blend of 12 exposures taken at 1/1600" to 1, and then stacked as a single "smart object," and later combined using a Mean stack mode. High-pass filters were also used to enhance coronal features. Taken through a 106 mm refractor equipped with a 0.85x reducer/flattener, at f/5, 500 mm focal length at ISO 100. Regulus is the bright star to the left of the Sun.

Of course, your camera may or may not dial in 1/800th of a second precisely, so you could then just round the exposure time down slightly to an even 1/1000th of a second, or maybe take a rapid series shots flipping between 1/500th, 1/750th, and 1/1000th of a second.

(Delve into eclipse photography with NASA's Photographing Eclipses: https://eclipse.gsfc.nasa.gov/SEhelp/SEphoto.html)

Apps such as Eclipse Orchestrator (http://www.moonglowtechnologies.com/products/EclipseOrchestrator/index.shtml) are also useful for giving you voice-commanded events through all phases of the eclipse. You can even use an app to automatically control the camera during totality, so you can just sit back and enjoy the experience.

PHOTOGRAPHING BRIGHT OBJECTS

The Moon is the very best astronomical target to cut your teeth on. Craters, mountains, and ridges jump right out at you, a real world awaiting photographic exploration. Plus, the Moon is bright enough to use standard daytime shooting methods. In fact, you might actually have to stop the aperture of the telescope down a bit near Full, as the Full Moon is about three to four times as bright as the 50-percent illuminated Quarter Moon.[9] One exception to the rule is during a total lunar eclipse: Be ready to rapidly dial down the exposure time from 1/500th of a second, to a second or more during totality.

ACTIVITY: BUILDING A HOMEMADE PLANETARY WEBCAM

Can't afford a DSLR? I've used cheap $20 webcams to image the Moon and planets for years. If you have a telescope, a laptop, and a webcam that you're willing to disassemble and modify for the cause, then you can start imaging planets, tonight. I used a similar rig for years to bring live images of the Moon, the planets, and brighter double stars into *Universe Today*'s Weekly Virtual Star Party.

The key modification is mounting the webcam to the telescope's eyepiece holder. Simply remove the lens off of the webcam, including any infrared filter (usually a thin meniscus on the sensor), then use cement glue to mount a standard 1¼-inch (3-cm) eyepiece barrel on the web cam lens opening.

In the field, you will need to control the camera manually, either using the camera's own software or a program such as K3CCD Tools. The sensor's field of view is tiny, not more than 10 to 15 arcminutes across, about half the size of a Full Moon. I like to center the target planet at high power, exchange the eyepiece for the webcam, then manually scale up the brightness and contrast of the camera until the out-of-focus, silvery doughnut of the planet appears. Then focus, center, and you're ready to start shooting.

K3CCD Tools will allow you to start processing stacks of images in the field, though for final processing such as wavelet stacking, contrast adjustment, and light sharpening, I prefer to use Registax and Photoshop.

K3CCD Tools: http://www.pk3.org/k3ccdtools/

Registax: https://www.astronomie.be/registax/

Another useful stacking program is PIPP: https://sites.google.com/site/astropipp/. PIPP is also a great pre-processing utility for cropping, centering, and preparing planetary images for stacking in a program such as Registax.

A simple setup for live webcasting or imaging the Moon and planets: A laptop, a modified webcam, and an 8" (20-cm) SCT telescope.

A tale of two modified, off-the-shelf webcams: one shooting through the telescope eyepiece afocally on a specialized mount (left) and the other with the lens removed, shooting directly through the telescope.

A fine view of the Pleiades (Messier 45) including the dust and nebulosity enshrouding the young open star cluster. This 14-hour image integration uses data obtained over two years to bring out an amazing degree of delicate detail.

DEEP SKY BEST TARGETS

The jump from shooting simple star trails and lunar and planetary shots to deep sky astrophotography is a big one. Deep sky targets are faint, meaning you'll need to track to take longer exposures.

What are some of the best first targets for astrophotography? Well, Chapter 5: The Sky From Season to Season (page 80) has a bevy of best targets for brighter deep sky astrophotography; we suggest the Andromeda Galaxy (M31), the Orion Nebula (M42), the Trifid Nebula (M20), and the open cluster (M35) in the constellation Gemini as good starter targets. Likewise, globular clusters such as Omega Centauri and M13 in Hercules are good dense, bright targets. I wouldn't start with tiny planetary nebulae or wispy emission nebulae, as these take extensive and precise tracking times, multiple filtered shots, and an extensive post-processing regimen.

The example of M31 is instructive in terms of focal length versus field of view.[10] Under suburban skies, you tend to merely see the bright core of this glorious extended object, which actually covers over 3 degrees. That's actually the width of six Full Moons, plus change.

On the next page, you'll find how the DSLR sensor size versus focal length affects the final field of view.

DSLR sensor size	Focal length	Field of View
	50	20 degrees
18 mm	200	5 degrees
	400	2.5 degrees
	50	26 degrees
24 mm	200	6.5 degrees
	400	3.3 degrees
	50	45 degrees
35 mm	200	10 degrees
	400	5 degrees

You can even capture tiny versions of other deep sky objects in the nearby sky, along with the Milky Way.

As you can see, a decent focal length and a full-frame, 35 mm DSLR is desirable, but by no means mandatory. Astronomy programs such as Sky Safari will also project the field of view of your selected equipment overlaying the sky, so you can plan which lenses and settings to use for selected targets.

SHOOTING DARK FRAMES

Noise in the form of "hot pixels" from camera defects or cosmic ray hits do occasionally turn up on time exposure images. An easy way to remove these is to create a dark frame image for subtraction. Do this during your imaging session for later subtraction post-processing, as the thermal sensitivity of your camera changes throughout the night as it cools down. A simple way to do this is to create one to twenty blank exposures at the same manual settings with the lens cap on.

You can even capture tiny versions of other deep sky objects in the nearby sky, along with the Milky Way.

Sirius, the brightest star in the sky with "star spikes."

GETTING ROUND VERSUS SPIKY STARS

When it comes to imaging, artifacts of light can have a strangely aesthetic appeal. CGI shots in movies, for example, still like to incorporate lens flares and perfect-looking internal reflections, often in shots where no true camera exists. Likewise, we've come to expect multi-spiked bright stars (known as **diffraction spikes**) in deep sky images, though ironically, we now have the technology to *erase* these artifacts from the image. If you're using a Newtonian reflector, for example, you already have a ready-made star-spike generator in the form of the spider-veins holding the secondary mirror in place. You can purchase aperture masks to make diffraction spikes . . . or you can simply make one, using nothing more than a collar and two strings or wire in a cross-hair configuration over the telescope or camera lens aperture. Thin gauge wire, fishing line, or light gauge guitar strings work great.

PROCESSING AND PROGRAMS

You could take college courses designed just around Photoshop. Some great quick fixes in Photoshop include the sharpening mask tool, contrast and brightness fixes, and a histogram map to show just where along the color scale an image peaks. Increasingly, astrophotographers are often now simply using Lightroom, importing the image to their smartphone to make quick fixes in the field with the free app.

Going with a black background often removes crucial details (e.g., wispy star lanes, faint galaxies and stars, airglow). You want a "neutral" versus "black" background.

"Don't be upset when your first images don't look like Hubble images," says Cory Schmitz (PhotographingSpace.com). "You will fail a lot, it's part of the learning curve. Pay attention to detail . . . the best way to do something is whatever gets you out under the stars the most."

Cory's formula for great astro-pics works as follows:

- Get as much light data as you can on your target.
- Shoot all the calibration frames (flats, darks, and bias) (but process even if you don't!). See Shooting Dark Frames on page 169.
- Reduce the data to increase the signal-to-noise ratio: stacking, bad frame removal, frame calibration, and so on in pre-processing.
- Post-process your reduced image with a proper software package to include histogram stretching, star reduction, saturation boosting, sharpening, curves and levels, and intelligent noise reduction as key parts of the workflow. The more precise you get, the better the result.
- Be sure to share those images!

As with cooking, the trick when it comes to image sharpening or histogram adjustment is knowing when just enough is enough to do the job. Good image processing is like good acting: you shouldn't notice it when it occurs. If a casual glance reveals image artifacts, you've probably pushed the processing too far. Dark cores and rings around stars, pixelation, and over-saturated planetary disks with bloated limbs are all sure signs of post-processing techniques taken to the extreme.

HONESTY IN ASTROPHOTOGRAPHY

Finally, be sure to be open and honest about disclosing all of the aspects of an image, including how it was shot and what equipment was used. If it's a mosaic or composite shot, say so; there's nothing wrong with making a composite image purely for artistic reasons, as long as it isn't passed off as anything otherwise. The camera reveals the beauty of the Universe that the eye is often insensitive to . . . hey, evolution only had water and jelly to work with, so it's amazing we can even see at all. With a focal length of 22 mm and a focal ratio of f/2.1,[11] you wouldn't even consider purchasing a camera as badly constructed as the eye for nighttime astrophotography. But all those splashy, vivid colors are indeed there, below the limits of perception.

And though members of the astrophotography community are a helpful bunch, always ready to give budding astrophotographers constructive criticism to help them along what's often a steep learning curve, they're also quick to spot fakes and blackball image fakers from the small online community.

A good composition for an astrophoto might tease out subtle detail, but still shows features that are really there, but are perhaps too faint to see. Adding to or manipulating the image—especially if it isn't documented as such—is a foul in many an astrophotographer's playbook.

. . . THERE'S NOTHING WRONG WITH MAKING A COMPOSITE IMAGE PURELY FOR ARTISTIC REASONS,

as long as it isn't passed off as anything otherwise.

A close grouping of the planets Mars, Venus, and the crescent Moon.

THE FUTURE OF ASTROPHOTOGRAPHY

MallinCams have also changed up the game, offering a near-live view of the deep sky with sensitive imagers. Another telescope project launched on Kickstarter by the SETI Institute and Unistellar for their Enhanced Vision Telescope may soon do the same, utilizing an electronic imager for real-time views. I think that there will always be room for the traditional approach of imaging the sky through a wide field of view, to truly capture the intricate wonders of the Universe.

Finally, don't forget to share those images with the astronomical community.

Sites to share your work:

Astronomy Picture of the Day (APOD): NASA's long-running website: https://apod.nasa.gov/apod/astropix.html

Space Weather: (frequently features amateur pics): http://www.spaceweather.com/

Instagram and Twitter. Most of the astrophotos in this book were shared by people on Instagram using our hashtag #UniverseToday.

The Universe Today Flickr pool: https://www.flickr.com/groups/universetoday/

References/suggested further reading: http://www.amazingsky.com/free-tutorials.html

TOP ASTRONOMY EVENTS FOR 2019–2024

A look at the top astronomical events over the next six years.

"I conclude that this star is not some kind of comet or fiery meteor . . . but that it is a star shining in the firmament itself—one that has never previously been seen before our time, in any age since the beginning of the world."

– Tycho Brahe, on the supernova of 1572

What's new under the Sun for the coming years?

The motion of the Moon and the planets are all predictable, and their paths through the sky provide an unfolding celestial spectacle that we can anticipate in advance. Annual meteor showers are also dependable, and while the next great comet could appear without warning at any time, short-term periodic comets come and go. And if astronomers are right, we might just be in for a Red Nova shining in Cygnus the Swan on or around 2022, though that probably *won't* rival Tycho's Star.

What follows is a brief rundown of what to expect out to the year 2024.

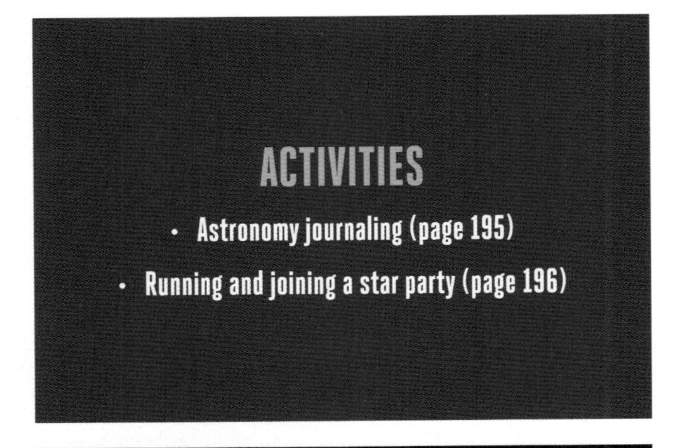

THE BIG IDEA

was to distill these events down to the very best

THE RULES

What we've constructed is a simple rundown of the top astronomical events by date. The big idea was to distill these events down to the very best, while providing a comprehensive overview for astronomical events for each year.

For the top events listed below for the entire year, we considered the following.

- Meteor showers with a ZHR greater than 10, where the phase of the Moon is not within a week of Full;
- Oppositions of the outer planets;
- Greatest elongations of the inner planets;
- Eclipses of the Sun and Moon;
- The closest conjunction of two naked eye planets and the best planet versus bright star event for the year;
- The best easily visible occultations of bright stars and planets for each year;
- Comets slated to reach perihelion in 2018 and forecast to break +10th magnitude.

Comets are always a tough call, as new bright comets come and go, often without warning. We included comets predicted to break +10th magnitude (binocular visibility) as listed by the British Astronomical Association (BAA). Naturally, the further out this look ahead is, the sparser the number of comets reaching favorable visibility becomes . . . binocular comets, and perhaps a score of good, new naked eye comets will no doubt come and go over the next few years, and the next great Comet of the Century could also arrive on the celestial scene at any time.

One caveat: Every year, when we look back at the big events in space news, we notice that most were actually surprises. You'll need to stay tuned to Universe Today to hear about those.

For occultations, we only included the very best ones involving bright stars (the big four of Aldebaran, Spica, Antares, and Regulus) and naked eye planets (sorry, Neptune and Uranus). Where the Moon will actually occult a bright star or planet, we noted so with the region from which it is visible, though these are also close conjunctions that are visible worldwide as well.

Lunar and solar eclipses, along with brief circumstances, are noted as well. For more information and the specific circumstances for every eclipse over a 5,000-year span from 2000 BC to 3000 AD, check out NASA's extensive eclipse site, listed in the resources section at the end of the book.

Fun with the white balance settings (daytime, tungsten, incandescent) to produce an artistic "Warhol Moon" mosaic.

A Full perigee "supermoon" rising over the Ericsson Globe Arena of Stockholm on May 5, 2012. Specs: taken with a DSLR at 200 mm, with a Canon 2x Extender. Exposure: 1/64" at f/8, ISO 400.

MOONS, BLACK AND BLUE

We also provided a brief table on what the Moon is up to for each year as well. This includes the following.

Blue Moons: For the intent of this guide, we used the modern definition of a Blue Moon as simply the second Full Moon, in a month with two.

Black Moons: Likewise, we defer to the definition of a Black Moon as the second New Moon in a month with two.

Supermoon: A modern term that has gained traction thanks to the internet, we note the Supermoon for each year as the Full Moon nearest to perigee timewise. Likewise, we refer to the Full Moon nearest to apogee as the Minimoon.

We also note the **Brown Lunation Number** for the start of each New Moon. This number counts synodic periods for the Moon going back to January 17, 1923.

Finally, note that '+/-' in respect to the Moon and its percentage of illumination is shorthand for waxing (+) and waning (-).

Each year gives a rundown of the very best of the best, astronomical events you shouldn't miss.

Ready? Let's see what's up for the next 6-year span.

2019

(The calendar dates are in bold; best events are underlined.)

The year 2019 contains a unique event: a rare 5-hour-and-29-minute transit of Mercury across the visible disc of the Sun, visible from Europe, Africa, the Middle East, and North/South America on November 11. This is only one of fourteen transits of Mercury for the twenty-first century. The only total solar eclipse for 2019 occurs on July 2, crossing the South American Andes mountains through Chile and Argentina, passing directly over several major astronomical observatories.

A profound solar minimum is also expected in 2019 through 2020, as the Sun transitions from Solar Cycle #24 to Cycle #25.

METEOR SHOWERS

One of the best meteor showers starts off 2019, as the brief <u>Quadrantids occur on **Jan 3**</u> with a ZHR=100, while the Moon is a 5 percent illuminated, waning crescent. Other favorable showers for 2019 include:

- The Eta Aquarids (**May 6**) ZHR=50, Moon 4 percent illuminated, waxing crescent
- The Daytime Arietids (**June 7**) ZHR=30, Moon 22 percent illuminated, waxing crescent
- The Orionids (**Oct 21**) ZHR=20, Moon 49 percent illuminated, waning crescent
- The Andromedids (**Dec 3**) ZHR=20, Moon 45 percent illuminated, waxing crescent
- The Ursids (**Dec 22**) ZHR=50, Moon 14 percent illuminated, waning crescent

THE OUTER PLANETS

Mars does not reach opposition in 2019, but instead spends most of the year low in the dusk sky, emerging into the early dawn in October. For the remainder of the outer planets, the best viewing seasons for each centered on opposition are:

- Jupiter (**June 10**)
- Saturn (**July 9**)
- Uranus (**Oct 28**)
- Neptune (**Sept 10**)
- Pluto (**Jul 14**)

THE INNER PLANETS

<u>Venus makes its best appearance early on in 2019, reaching maximum elongation on **Jan 6**</u> 47 degrees west (dawn) of the Sun. Mercury reaches greatest elongation six times in 2019:

- Mercury (**Feb 27**), 18 degrees east (dusk)
- Mercury (**Apr 11**), 28 degrees west (dawn)
- Mercury (**Jun 23**), 25 degrees east (dusk)
- Mercury (**Aug 9**), 19 degrees west (dawn)
- Mercury (**Oct 20**), 25 degrees east (dusk)
- Mercury (**Nov 28**), 20 degrees west (dawn)

Mercury Transit of November 11, 2019

Geocentric Diagram and Visibility Map

Geometric Transit: 15:19:47.4 UT J.D.:2458799.138743

Constants
△T: 69.70s

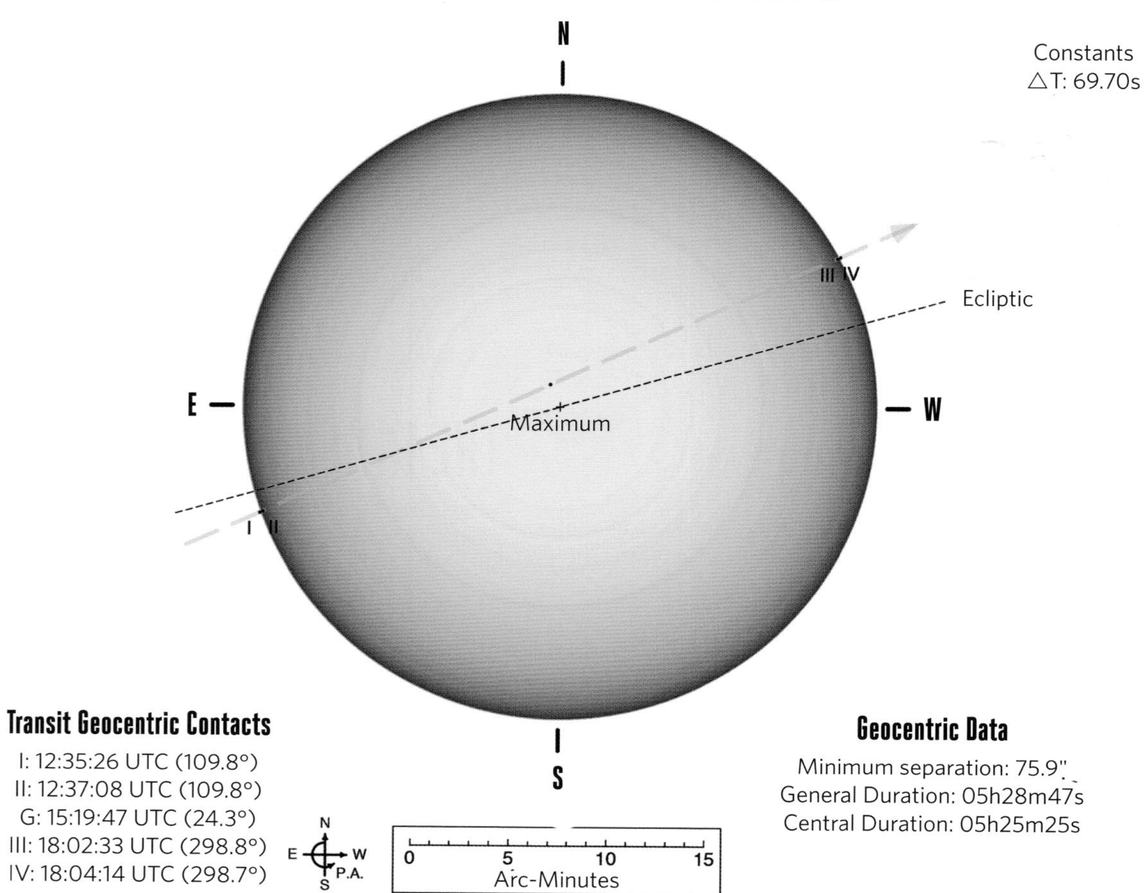

Transit Geocentric Contacts

I: 12:35:26 UTC (109.8°)
II: 12:37:08 UTC (109.8°)
G: 15:19:47 UTC (24.3°)
III: 18:02:33 UTC (298.8°)
IV: 18:04:14 UTC (298.7°)

N
E W
S P.A.

0 5 10 15
Arc-Minutes

Geocentric Data

Minimum separation: 75.9"
General Duration: 05h28m47s
Central Duration: 05h25m25s

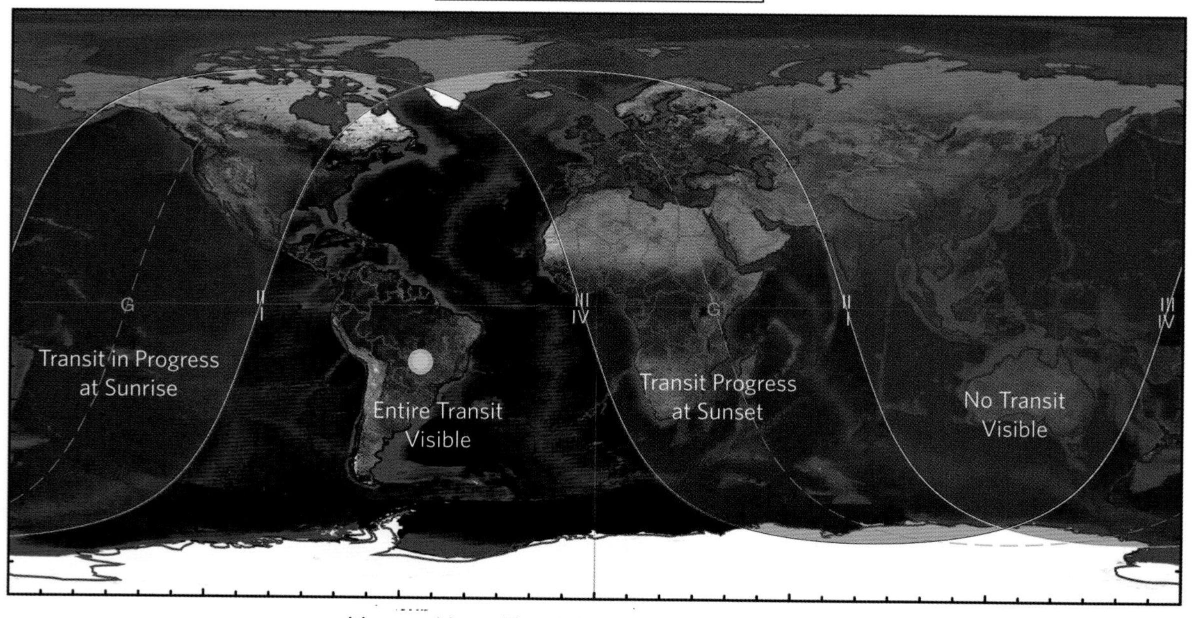

Mercury Venus Transit Maestro - Xavier M. Jubier
(http://xjubier.free.fr/)

ECLIPSES

There are five eclipses in 2019, three solar and two lunar:

- A partial solar eclipse **(Jan 6)** maximum partial phase of 71 percent, favoring Northeast Asia and the Northern Pacific.
- A total lunar eclipse **(Jan 21)** totality duration 1 hour, 1 minute favoring the Atlantic, North and South America.
- Solar total eclipse **(Jul 2)**, (maximum totality of 4 minutes, 33 seconds) covering the Southern Pacific and Chile/Argentina.
- A partial lunar eclipse **(Jul 16)** maximum partial phase of 65 percent, favoring Africa, Europe, Southern Asia, and Australia.
- An annular solar eclipse **(Dec 26)** maximum annularity of 3 minutes, 40 seconds covering Southeast Asia.

TRANSITS

Mercury transits the disk of the Sun on **(Nov 11)**.

CLOSEST CONJUNCTIONS

- Planet-planet: Mercury-Mars **(June 18)**, 12 arcminutes apart, 24 degrees from the Sun in the dusk
- Planet-star: Mars-Regulus **(Aug 18)**, 38 arcminutes apart, 5 degrees from the Sun in the dusk

OCCULTATIONS OF PLANETS

There are twenty occultations involving the Moon and five planets in 2019:

- Mercury (1): **(Feb 5)**, Southeast Asia, +0.2 percent Moon
- Venus (2): **(Jan 31)**, Central Pacific, -18 percent Moon; **(Dec 29)**, Antarctica, +10 percent Moon
- Mars (1): **(Jul 4)**, Central Pacific, -3 percent Moon
- Jupiter (2): **(Nov 28)**, South-central Asia, +3 percent Moon; **(Dec 26)**, Southeast Asia, +0.1 percent Moon
- Saturn (13): **(Jan 5)**, North America, -0.1 percent Moon; **(Feb 2)**, North Africa, -5.7 percent Moon; **(Mar 1)**, Central Pacific, -23 percent Moon; **(Mar 29)**, Southern Africa, -39 percent Moon; **(Apr 25)**, New Zealand, Southeast Australia, -66 percent Moon; **(May 22)**, South Africa, -88 percent Moon; **(Jun 19)**, southern South America, -96 percent Moon; **(Jul 16)**, South America, +99 percent Moon; **(Aug 12)**, Australia South Pacific, 91 percent Moon; **(Sep 8)**, Southeast Asia, +72 percent Moon; **(Oct 5)**, South Africa, +46 percent Moon; **(Nov 2)**, New Zealand, +30 percent Moon; **(Nov 29)**, South Atlantic, +9 percent Moon; **(Dec 27)**, Antarctica, +1 percent Moon
- Best occultation (star): The Moon does not occult a +1 magnitude or brighter star in 2019.

Map showing the November 11, 2019, transit of Mercury across the disc of the Sun. The top graphic shows the chord of the 5-hour-and-28-minute transit, while the bottom map shows where the transit is visible. South America and the Eastern United States see the transit in its entirety, while the rest of North America sees the transit in progress at sunrise, and Africa and Europe see the transit in progress at sunset. Only Australia and the Far East miss out on the transit entirely.

MOON PHASES FOR 2019 (IN UNIVERSAL TIME)			
Closest Perigee: Feb 19 (356,761 km)		Most-Distant Apogee: Feb 5 (406,555 km)	
New Moon	Brown Lunation Number	Full Moon	Notes
Jan 6 - 1:30 UT	1188	Jan 21 - 5:17 UT	Eclipse season
Feb 4 - 21:05 UT	1189	Feb 19 - 15:54 UT	Super Full Moon
Mar 6 - 16:05 UT	1190	Mar 21 - 1:43 UT	
Apr 5 - 8:52 UT	1191	Apr 19 - 11:12 UT	
May 4 - 22:47 UT	1192	May 18 - 21:12 UT	
Jun 3 - 10:03 UT	1193	Jun 17 - 8:31 UT	
Jul 2 - 19:17 UT	1194	Jul 16 - 21:40 UT	Eclipse season
Aug 1 - 3:13 UT	1195	Aug 15 - 12:31 UT	
Aug 30 - 10:38 UT	1196-Black Moon (New)	Sep 14 - 4:35 UT	Minimoon (Full)
Sep 28 - 18:28 UT	1197	Oct 13 - 21:11 UT	
Oct 28 - 3:40 UT	1198	Nov 12 - 13:37 UT	
Nov 26 - 15:08 UT	1199	Dec 12 - 5:15 UT	
Dec 26 - 5:16 UT	1200		Eclipse season

PERIODIC COMETS OVER MAGNITUDE +10

- 289P/Blanpain (perihelion **Dec 21**, magnitude +6?) discovered in 1819, will only be favorable if it provides a surprise outburst.
- 322P/SOHO (perihelion **Aug 31**, magnitude +7). 322P is SOHO's first periodic comet discovery.

2020

The planet Venus rules 2020, with one each favorable apparitions in the dawn and dusk. Mars also makes a fine apparition in October, reaching 22 arcseconds in size. Watch for a dramatic occultation of Mars by the waning crescent Moon on the morning of February 18. A fine close pairing of Venus and Jupiter in the dusk also ends out the year.

The solar minimum marking the end of Cycle #24 and the start of Cycle #25 should also occur on or around mid-2020 . . . or will solar cycle #25 be absent altogether?

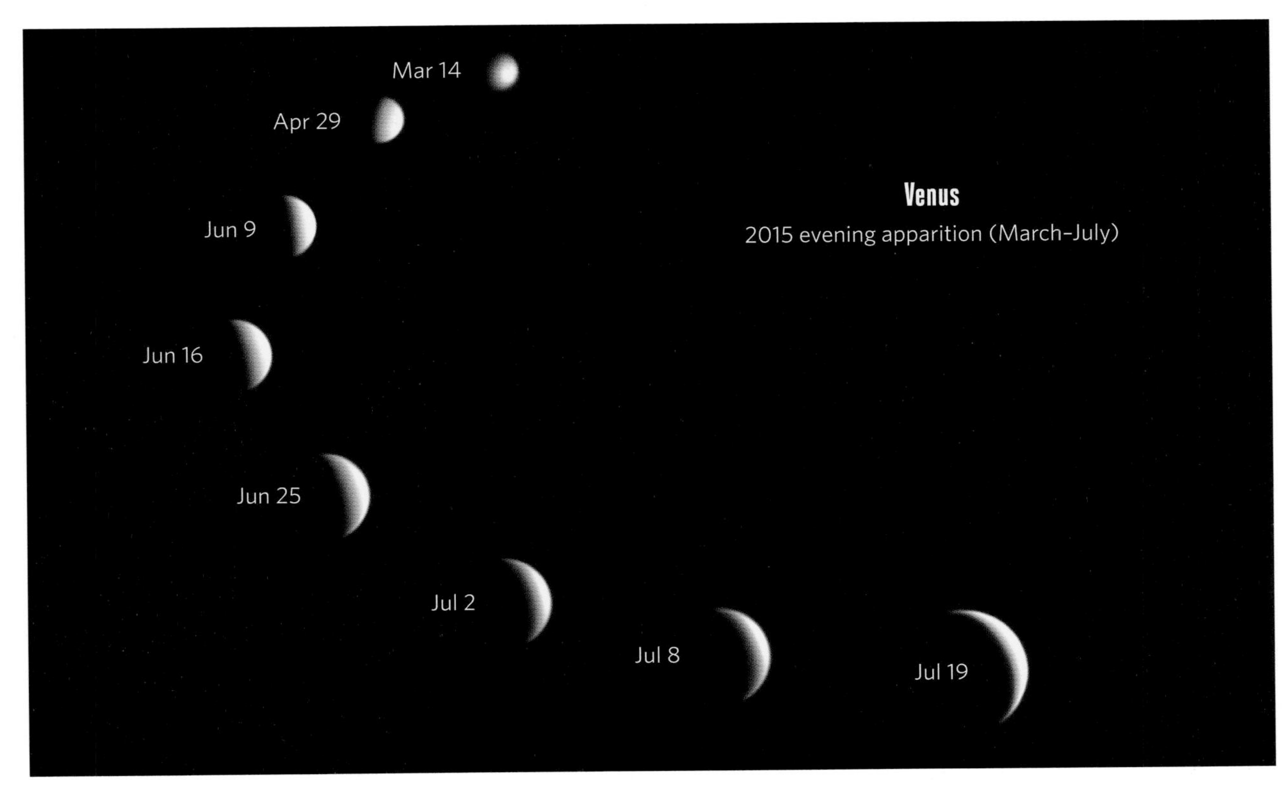

A phase sequence of the planet Venus through 2015, from near Full phase to approaching inferior conjunction as a thin crescent.

METEOR SHOWERS

Two top meteor showers to watch for in 2020 are the surefire August Perseids and the December Geminids. Other favorable showers for the year include:

- The Lyrids (**Apr 22**) ZHR=20, Moon 0.5 percent illuminated, waning crescent
- The June Boötids (**Jun 27**), ZHR=50, Moon 41 percent illuminated, waxing crescent
- The Perseids (**Aug 12**), ZHR=100, Moon 40 percent illuminated, waning crescent
- Taurids (**Oct 10**), ZHR=10, Moon 43 percent illuminated, waning crescent
- The Orionids (**Oct 21**), ZHR=20, Moon 32 percent illuminated, waxing crescent
- The Leonids (**Nov 17**), ZHR=20, Moon 10 percent illuminated, waxing crescent
- The Geminids (**Dec 14**), ZHR=120, Moon 0.1 percent illuminated, New Moon

THE OUTER PLANETS

Oppositions centered on the best observing seasons for each of the naked eye outer planets in 2020 are:

- Mars (**Oct 13**)
- Jupiter (**Jul 14**)
- Saturn (**Jul 20**)
- Uranus (**Oct 31**)
- Neptune (**Sep 11**)
- Pluto (**Jul 15**)

THE INNER PLANETS

Venus reaches greatest elongation twice in 2020:

- Venus (**Mar 24**) 46 degrees east (dusk)
- Venus (**Aug 13**) 46 degrees west (dawn)

Mercury reaches greatest elongation six times in 2020:

- Mercury (**Feb 10**) 18 degrees east (dusk)
- Mercury (**Mar 24**) 28 degrees west (dawn)
- Mercury (**Jun 4**) 24 degrees east (dusk)
- Mercury (**Jul 22**) 20 degrees west (dawn)
- Mercury (**Oct 1**) 26 degrees east (dusk)
- Mercury (**Nov 10**) 19 degrees west (dawn)

ECLIPSES

There are six eclipses in 2020, two solar and four lunar:

- A penumbral lunar eclipse (**Jan 10**), with a maximum penumbral immersion of 88 percent for Europe, Africa, Asia, and Australia.
- A penumbral lunar eclipse (**Jun 5**), with a maximum penumbral immersion of 59 percent for Europe, Africa, Asia, and Australia.
- An annular solar eclipse (**Jun 21**), with a maximum duration of annularity of 38 seconds crossing central Africa, Southern Asia, China, and the Pacific.
- A penumbral lunar eclipse (**Jul 5**), with a maximum penumbral immersion of 36 percent for the Americas, Southwest Europe, and Africa.
- A penumbral lunar eclipse (**Nov 30**), with a maximum penumbral immersion of 74 percent for Asia, Australia, the Pacific, and the Americas.
- A total solar eclipse (**Dec 14**), with a maximum duration of 2 minutes and 10 seconds, with totality crossing the South Pacific, Chile, Argentina, and the South Atlantic.

CLOSEST CONJUNCTIONS

- Planet-planet: Jupiter-Saturn (**Dec 21**), 6 arcminutes apart, 30 degrees east of the Sun in the dusk. This is a fine close conjunction, with the two gas giants passing ⅕th the apparent distance apart of a Full Moon.
- Planet-star: Venus-Regulus (**Oct 2**), 5 arcminutes apart, 40 degrees west of the Sun in the dawn.

OCCULTATIONS OF PLANETS

- Best occultation (planet): Mars for North America on (**Feb 18**). There are ten occultations in 2020 involving the Moon and four planets:
- Mercury (1): (**Dec 14**) Europe, 0.1 percent Moon (unobservable, only 3 degrees from the Sun)
- Venus (2): (**June 19**) Northeast North America, -4 percent Moon; (Dec 12) Northeast Asia, -5 percent Moon
- Mars (5): (**Feb 18**) North America, -24 percent Moon; (**Mar 18**) the southernmost tip of South America, -30 percent Moon; (**Aug 9**) Southeast South America, -72 percent Moon; (**Sep 6**) central South America, -85 percent Moon; (**Oct 3**) the southernmost tip of South America, -98 percent Moon
- Jupiter (2): (**Jan 23**) Australia/New Zealand in the daytime, -3 percent Moon; (**Feb 19**) Antarctica, -13 percent Moon
- Best occultation (star): The Moon does not occult a +1 magnitude or brighter star in 2020.

PERIODIC COMETS OVER MAGNITUDE +10

- P/2017 T2 PanSTARRS (perihelion **May 6**, may reach +8 magnitude) as it crosses into Ursa Major
- 88/P Howell (perihelion **Sept 26**, magnitude +9) near the Libra-Scorpius border
- 2/P Encke reaches (perihelion **Jun 26**, magnitude +8) in Gemini

MOON PHASES FOR 2020 (IN UNIVERSAL TIME)			
Closest Perigee: Apr 7 (356,908 km)		Most-Distant Apogee: Mar 24 (406,688 km)	
New Moon	Brown Lunation Number	Full Moon	Notes
	1200	Jan 10 - 19:23 UT	Eclipse season
Jan 24 - 21:44 UT	1201	Feb 9 - 7:35 UT	
Feb 23 - 15:34 UT	1202	Mar 9 - 17:49 UT	
Mar 24 - 9:30 UT	1203	Apr 8 - 2:36 UT	Supermoon
Apr 23 - 2:27 UT	1204	May 7 - 10:46 UT	
May 22 - 17:40 UT	1205	Jun 5 - 19:13 UT	
Jun 21 - 6:42 UT	1206	Jul 5 - 4:45 UT	Eclipse season
Jul 20 - 17:34 UT	1207	Aug 3 - 16:00 UT	
Aug 19 - 2:42 UT	1208	Sep 2 - 5:23 UT	
Sep 17 - 11:01 UT	1209	Oct 1 - 21:07 UT	
Oct 16 - 19:32 UT	1210	Oct 31 - 14:51 UT	Blue Moon, Minimoon
Nov 15 - 5:09 UT	1211	Nov 30 - 9:32 UT	
Dec 14 - 16:19 UT	1212	Dec 30 - 3:30 UT	Eclipse season

2021

The year 2021 features a high Arctic annular and an Antarctic total solar eclipse, a fine total lunar eclipse centered on the Pacific Rim region, and a dramatic occultation of Mars in the dusk sky by the waxing crescent Moon for Southeast Asia. The year also features an expected handful of periodic binocular comets.

Solar Cycle #25 should be underway in 2021, though if the sputtering start of Cycle #24 a decade ago was any indication, the next sunspot cycle could be lackluster at best.

METEOR SHOWERS

The year 2021 is a great year for that summer standby: the Perseid meteors. The Fall Taurid fireballs should also be enhanced in 2020–2021, owing to the recent perihelion passage of their source, Comet 2P/Encke.

- Eta Aquarids (**May 6**), ZHR=50, Moon 23 percent illuminated, waning crescent
- Daytime Arietids (**Jun 7**), ZHR=30, Moon 7 percent illuminated, waning crescent
- Perseids (**Aug 12**), ZHR=100, Moon 18 percent illuminated, waxing crescent
- Taurids (**Oct 10**), ZHR=10, Moon 22 percent illuminated, waxing crescent
- Andromedids (**Dec 3**), ZHR=20, Moon 1 percent illuminated, waning crescent

THE OUTER PLANETS

The planet Mars does not reach opposition in 2021, but instead spends the start of 2021 loitering low in the dusk sky before slowly reemerging in the dawn after February. The best observing seasons for the remainder of the outer planets centered on opposition are:

- Jupiter (**Aug 20**)
- Saturn (**Aug 2**)
- Uranus (**Nov 4**)
- Neptune (**Sep 14**)
- Pluto (**Jul 17**)

THE INNER PLANETS

The planet Venus rules the dusk sky in 2021, reaching greatest dusk elongation on **Oct 29**, 47 degrees east of the Sun.

Mercury reaches greatest elongation six times in 2021:

- Mercury (**Jan 24**), 19 degrees east (dusk)
- Mercury (**Mar 6**), 27 degrees west (dawn)
- Mercury (**May 17**), 22 degrees east (dusk)
- Mercury (**Jul 4**), 22 degrees west (dawn)
- Mercury (**Sep 14**), 27 degrees east (dusk)
- Mercury (**Oct 25**), 18 degrees west (dawn)

ECLIPSES

There are four eclipses in 2021, two solar and two lunar:

- A total lunar eclipse (**May 26**) with a maximum duration of 15 minutes centered on the Pacific Rim region.
- An annular solar eclipse (**Jun 10**), maximum duration of annularity is 3 minutes, 51 seconds crossing through the Arctic.
- A partial lunar eclipse (**Nov 19**) with a maximum partial phase of 97 percent, favoring the Americas, Northern Europe, East Asia, Australia, and the Pacific.
- A total solar eclipse (**Dec 4**) with a maximum length for totality of 1 minute, 54 seconds crossing the Antarctic.

CLOSEST CONJUNCTIONS

- Planet-planet: Mercury–Mars (**Aug 18**), 4 arcminutes apart, 17 degrees east of the Sun in the dusk sky.
- Planet-star: Mars–Regulus (**Jul 30**), 36 arcminutes apart, 23 degrees east of the Sun in the dusk sky.

OCCULTATIONS OF PLANETS

There are seven planetary occultations in 2021 by the Moon involving three planets:

- Mercury (2): (**Nov 3**), northeast North America, -2 percent illuminated Moon; (**Dec 4**), South America, South Africa, +1 percent Moon
- Venus (2): (**May 12**), South Pacific, +1 percent illuminated Moon; (**Nov 8**), northwest Pacific, +20 percent illuminated Moon
- Mars (3): (**Apr 17**), Southeast Asia, +26 percent illuminated Moon; (**Dec 3**), Northeast Asia, -1 percent illuminated Moon; (**Dec 31**), Southeast Asia, -6 percent illuminated Moon.
- Best occultation (star): The Moon does not occult a star brighter than +1st magnitude in 2021.

PERIODIC COMETS OVER MAGNITUDE +10

A handful of periodic comets attain binocular visibility in 2021:

- 7P/Pons-Winnecke (perihelion **May 26**, magnitude +8) in the constellation Aquarius
- 15P/Finlay (perihelion **Jul 13**, magnitude +9) in the constellation Taurus
- 8P/Tuttle (perihelion **Aug 21**, magnitude +9) in the constellation Cancer
- 6P/d'Arrest (perihelion **Sep 17**, magnitude +9) in the constellation Sagittarius

The three stages of a total lunar eclipse, from partial, through totality, and back to partial phase again. Taken during the January 31, 2018, total lunar eclipse. This is a combination of shots taken from two cameras, the 1st DSLR: 14mm lens at f/2.8 and various ISO settings for the foreground; the 2nd DSLR: zoomed in 120mm to focus on the eclipsed Moon, set at various ISOs and shutter speeds through the eclipse.

MOON PHASES FOR 2021 (IN UNIVERSAL TIME)			
Closest Perigee: May 26 (357,309 km)		Most-Distant Apogee: May 11 (406,511 km)	
New Moon	Brown Lunation Number	Full Moon	Notes
Jan 13 - 5:03 UT	1213	Jan 28 - 19:19 UT	
Feb 11 - 19:08 UT	1214	Feb 27 - 8:20 UT	
Mar 13 - 10:24 UT	1215	Mar 28 - 18:50 UT	
Apr 12 - 2:33 UT	1216	Apr 27 - 3:33 UT	
May 11 - 19:02 UT	1217-Minimoon	May 26 - 11:15 UT	Eclipse season/Supermoon
Jun 10 - 10:54 UT	1218	Jun 24 - 18:40 UT	
Jul 10 - 1:18 UT	1219	Jul 24 - 2:37 UT	
Aug 8 - 13:51 UT	1220	Aug 22 - 12:02 UT	
Sep 7 - 0:52 UT	1221	Sep 20 - 23:55 UT	
Oct 6 - 11:06 UT	1222	Oct 20 - 14:58 UT	
Nov 4 - 21:15 UT	1223	Nov 19 - 9:00 UT	Eclipse season
Dec 4 - 7:45 UT	1224	Dec 19 - 4:38 UT	

TOP ASTRONOMY EVENTS FOR 2022

The year 2022 is a fine year for observing planets, with a favorable opposition of Mars, while Venus rules the dawn along with all of the other classical, naked eye planets in the first few months of the year. It also sees two total lunar eclipses visible from the Americas, and a good occultation of Mars on December 8 for North America. And will Red Nova KIC 9832227 pop this year?[1] Solar activity for Cycle #25 should ramp up in 2022, meaning more sunspots and increased auroral activity.

A stunning image of a gas jet from Comet 67/P Churyumov-Gerasimenko captured by the European Space Ageny's Rosetta mission.

METEOR SHOWERS

The year 2022 is an off year for surefire favorites such as the Perseids and Geminids, letting lesser-known annual meteor showers such as the Ursids and Delta Aquarids shine. It may also be an excellent year for the June Boötids, as the shower's parent comet 7P Pons-Winnecke reached perihelion just last year.

- Quadrantids (**Jan 3**), ZHR=60, Moon 1 percent illuminated, waxing crescent
- Eta Aquarids (**May 6**), ZHR=50, Moon 27 percent illuminated, waxing crescent
- June Boötids (**Jun 27**), ZHR=50, Moon 2 percent illuminated, waning crescent
- Delta Aquarids (**Jul 30**), ZHR=30, Moon 3 percent illuminated, waxing crescent
- Orionids (**Oct 21**), ZHR=20, Moon 16 percent illuminated, waning crescent
- Leonids (**Nov 17**), ZHR=20, Moon 40 percent illuminated, waning crescent
- Ursids (**Dec 22**), ZHR=50, Moon 1 percent illuminated, waning crescent

THE OUTER PLANETS

The year 2022 is a Mars Year, as the Red Planet reaches opposition (**Dec 8**) in the astronomical constellation of Taurus the Bull, displaying a disk 17 arcseconds across. For the remainder of the outer planets, the best observing seasons centered on opposition are:

- Jupiter (**Sep 26**)
- Saturn (**Aug 14**)
- Uranus (**Nov 9**)
- Neptune (**Sep 16**)
- Pluto (**Jul 20**)

THE INNER PLANETS

Venus rules the dawn in the first half of 2022, with the company of the other naked eye planets in the northern hemisphere spring, culminating with a greatest elongation of 47 degrees west (dawn) on **Mar 20**. Mercury reaches greatest elongation seven times in 2022:

- Mercury (**Jan 7**) 19 degrees east (dusk)
- Mercury (**Feb 16**) 26 degrees west (dawn)
- Mercury (**Apr 29**) 21 degrees east (dusk)
- Mercury (**Jun 16**) 23 degrees west (dawn)
- Mercury (**Aug 27**) 27 degrees east (dusk)
- Mercury (**Oct 8**) 18 degrees west (dawn)
- Mercury (**Dec 21**) 20 degrees east (dusk)

ECLIPSES

There are a minimal number of eclipses in 2022, two solar and two lunar:

- A partial solar eclipse (**Apr 30**) with a maximum partial phase of 64 percent, favoring the southeastern Pacific and southernmost South America.
- A total lunar eclipse (**May 16**) with totality lasting 1 hour, 25 minutes, favoring the Americas, Europe, and Africa.
- A partial solar eclipse (**Oct 25**) with a maximum partial phase of 86 percent, favoring Europe, northeast Africa, the Middle East, and Western Asia.
- A total lunar eclipse (**Nov 8**) with totality lasting 1 hour, 25 minutes, favoring Asia, Australia, the Pacific, and the Americas.

CLOSEST CONJUNCTIONS

- Planet-planet: Venus-Jupiter (**Apr 30**), 13 arcminutes apart, 42 degrees west of the Sun in the dawn sky.
- Planet-star: Venus-Regulus (**Sep 5**), 42 arcminutes apart, 13 degrees west of the Sun in the dawn sky.

OCCULTATIONS OF PLANETS

There are seven occultations involving the Moon and three planets in 2022:

- Mercury (2): (**Oct 24**), northwest North America, -1 percent Moon; (**Nov 24**), Antarctica, +1 percent Moon
- Venus (2): (**May 27**), Madagascar, -11 percent Moon; (**Oct 25**), South Africa (daytime), 0 percent Moon (unobservable, shortly after partial solar eclipse)
- Mars (3): (**Jun 22**), Antarctica, -33 percent Moon; (**Jul 21**), northeast Asia, -39 percent Moon; (**Dec 8**), North America and Europe, 99.9 percent Full Moon (on the same date as opposition for Mars)
- Best occultation (star): The Moon does not occult a +1 magnitude or brighter star in 2022.

PERIODIC COMETS OVER MAGNITUDE +10

67P/Churyumov-Gerasimenko was visited by the European Space Agency's Rosetta spacecraft and Philae lander in 2014–2016, and will also reach binocular visibility in 2022. If approved, NASA's CAESAR sample return mission will revisit Comet 67P for a sample return to Earth in 2038. Comet C/2017 K2 PanSTARRS may also break naked eye brightness in 2022.

Here are the top visible comets expected in 2022:

- 67P/Churyumov-Gerasimenko (perihelion **Jan 22**, magnitude +8) in the constellation Cancer
- 19P/Borrelly (perihelion **Feb 1**, magnitude +8) in the constellation Pisces
- C/2017 K2 PanSTARRS (perihelion **Dec 19**, magnitude +6) in the constellation Ara
- 118P/Shoemaker-Levy (perihelion **Nov 24**, magnitude +9) in the constellation Cancer

MOON PHASES FOR 2022 (IN UNIVERSAL TIME)			
Closest Perigee: Jul 13 (357,263 km)		Most-Distant Apogee: Jun 29 (406,580 km)	
New Moon	Brown Lunation Number	Full Moon	Notes
Jan 2 - 18:36 UT	1225	Jan 17 - 23:51 UT	
Feb 1 - 5:49 UT	1226	Feb 16 - 17:00 UT	
Mar 2 - 17:38 UT	1227	Mar 18 - 7:21 UT	
Apr 1 - 6:28 UT	1228	Apr 16 - 18:58 UT	
Apr 30 - 20:31 UT	1229 - Black Moon	May 16 - 4:16 UT	Eclipse season
May 30 - 11:32 UT	1230	Jun 14 - 11:53 UT	
Jun 29 - 2:53 UT	1231	Jul 13 - 18:38 UT	
Jul 28 - 17:55 UT	1232	Aug 12 - 1:36 UT	
Aug 27 - 8:17 UT	1233	Sep 10 - 9:59 UT	
Sep 25 - 21:54 UT	1234	Oct 9 - 20:55 UT	
Oct 25 - 10:48 UT	1235	Nov 8 - 11:03 UT	Eclipse season
Nov 23 - 22:57 UT	1236	Dec 8 - 4:10 UT	
Dec 23 - 10:18 UT	1237		

2023

The planet Mars continues to shine brightly in the evening sky in early 2023, fresh off of its December 2022 opposition. Watch for a fine occultation of Mars by the Moon for North America on January 31. A rare hybrid eclipse also crosses Indonesia, and an annular solar eclipse crosses North and South America in 2023. This is also a good year for the Perseids and Geminid meteor showers, and the Leonids may begin to show enhanced activity toward their 2032–2033 peak in the upcoming decade.

METEOR SHOWERS

The year 2023 is a great year for those annual standbys, the August Perseids and the December Geminids. Other favorable meteor showers in 2023 include:

- The Lyrids (**Apr 22**), ZHR=20, Moon 5 percent illuminated, waxing crescent
- The Perseids (**Aug 12**), ZHR=100, Moon 15 percent illuminated, waning crescent
- The Taurids (**Oct 10**), ZHR=10, Moon 17 percent illuminated, waning crescent
- The Orionids (**Oct 21**), ZHR=20, Moon 40 percent illuminated, waxing crescent
- The Leonids (**Nov 17**), ZHR=20, Moon 17 percent illuminated, waxing crescent
- The Geminids (**Dec 14**), ZHR=120, Moon 2 percent illuminated, waxing crescent

THE OUTER PLANETS

Mars, coming off an opposition in December 2022, recedes from the Earth in 2023. The best seasons to observe the other outer planets centered on their respective opposition dates are:

- Jupiter (**Nov 3**)
- Saturn (**Aug 27**)
- Uranus (**Nov 13**)
- Neptune (**Sep 19**)
- Pluto (**Jul 22**)

THE INNER PLANETS

Venus reaches greatest elongation twice in 2023: once on Jun 4, 45 degrees east of the Sun in the dusk, and on **Oct 23**, 46 degrees west of the Sun in the dawn sky.

The planet Mercury reaches greatest elongation six times in 2023:

- Mercury (**Jan 30**) 25 degrees west (dawn)
- Mercury (**Apr 11**) 20 degrees east (dusk)
- Mercury (**May 29**) 25 degrees west (dawn)
- Mercury (**Aug 10**) 27 degrees east (dusk)
- Mercury (**Sep 22**) 18 degrees west (dawn)
- Mercury (**Dec 4**) 21 degrees east (dusk)

ECLIPSES

The year 2023 offers four eclipses: two lunar, and two solar, including a rare hybrid solar eclipse which is annular along part of the track and total along another.

- A hybrid solar eclipse (**Apr 20**), with a maximum totality of 1 minute, 16 seconds crossing Indonesia, Australia, and Papua New Guinea.
- A penumbral lunar eclipse (**May 5**), with the Moon 96 percent immersed in the Earth's penumbral shadow favoring Africa, Asia, and Australia.
- An annular solar eclipse (**Oct 14**), with a maximum annularity of 5 minutes, 17 seconds crossing the western United States, Central America, Colombia, and Brazil.
- A partial lunar eclipse (**Oct 28**), with a maximum of 12 percent of the Moon eclipsed by the Earth's umbra favoring the eastern Americas, Europe, Africa, Asia, and Australia.

CLOSEST CONJUNCTIONS

- Planet-planet: Venus-Saturn (**Jan 22**), 20 arcminutes apart, 22 east of the Sun in the dusk sky.
- Planet-star: Mercury-Regulus (**Jul 29**), 6 arcminutes apart, 25 degrees east of the Sun in the dusk sky.

An image sequence showing the stages of the May 20, 2012, annular solar eclipse from Albuquerque, New Mexico.

OCCULTATIONS

- Best occultation (planet): There are twelve occultations involving the Moon and four planets in 2023:
  - Mercury (1): (**Oct 14**), South Africa (daytime), -0.1 percent Moon (unobservable)
  - Venus (2): (**Mar 24**), Southeast Asia, +9 percent Moon; (**Nov 9**), Greenland, -15 percent Moon
  - Mars (5): (**Jan 3**), Southern Africa, +92 percent Moon; (**Jan 31**), Southern U.S./Mexico, +74 percent Moon; (**Feb 28**), Iceland/northern Scandinavia, +59 percent Moon; (**Sep 16**), northeast South America (North America in the daytime), +3 percent Moon; (**Oct 15**), Antarctica, +1 percent Moon
  - Jupiter (4): (**Feb 22**), Southern South America, +10 percent Moon; (**Mar 22**), Eastern Caribbean, +2 percent Moon; (**Apr 19**), Central America/ Caribbean, -0.2 percent Moon (unobservable)
- Best occultation (star): The Moon occults Antares five times in 2023:
  - Antares: (**Aug 25**), North America, +58 percent Moon
  - Antares: (**Sep 21**), western Pacific, +35 percent Moon
  - Antares: (**Oct 18**), the Middle East, +15 percent Moon
  - Antares: (**Nov 14**), eastern North America, +3 percent Moon
  - Antares: (**Dec 12**), Southeast Asia (daytime), -1 percent Moon

PERIODIC COMETS OVER MAGNITUDE +10

As of writing this in 2017, no comets are set to break +10 magnitude, though that will change as we get closer to 2023.

MOON PHASES FOR 2023 (IN UNIVERSAL TIME)			
Closest Perigee: Jan 21 (356,569 km)		Most-Distant Apogee: Feb 4 (406,475 km)	
New Moon	Brown Lunation Number	Full Moon	Notes
		Jan 6 - 23:10 UT	
Jan 21 - 20:56 UT	1238	Feb 5 - 18:31 UT	Minimoon
Feb 20 - 7:09 UT	1239	Mar 7 - 12:43 UT	
Mar 21 - 17:27 UT	1240	Apr 6 - 4:37 UT	Eclipse season
Apr 20 - 4:16 UT	1241	May 5 - 17:37 UT	
May 19 - 15:56 UT	1242	Jun 4 - 3:44 UT	
Jun 18 - 4:39 UT	1243	Jul 3 - 11:41 UT	
Jul 17 - 18:34 UT	1244	Aug 1 - 18:33 UT	
Aug 16 - 9:39 UT	1245	Aug 31 - 1:37 UT	Blue/Supermoon
Sep 15 - 1:40 UT	1246	Sep 29 - 9:58 UT	Eclipse season
Oct 14 - 17:55 UT	1247	Oct 28 - 20:24 UT	
Nov 13 - 9:27 UT	1248	Nov 27 - 9:17 UT	
Dec 12 - 23:32 UT	1249	Dec 27 - 00:34 UT	

2024

It's here. Just under 7 years after the Great American Eclipse of August 21, 2017, the Moon's shadow crosses the United States once again on April 8, 2024. This time, however, Mexico and the Canadian Maritime provinces get to join in on the show as well. Crossing the U.S. from southwest to northeast, this path also crisscrosses the 2017 path centered on the tristate Illinois, Kentucky, and Missouri region. This eclipse will be the big-ticket astronomical event for 2024. This eclipse also has a personal significance to me, as the path passes right over my hometown of Mapleton, Maine . . . I think I know where I'll be!

METEOR SHOWERS

The year 2024 is an off year for most of the larger annual meteor showers, leaving the Aquarid streams to lead the year:

- Eta Aquarids (**May 6**), ZHR=50, Moon 1 percent illuminated, waning crescent
- Daytime Arietids (**Jun 7**), ZHR=30, Moon 4 percent illuminated, waxing crescent
- Delta Aquarids (**Jul 30**), ZHR=30, Moon 18 percent illuminated, waning crescent
- Taurids (**Oct 10**), ZHR=10, Moon 45 percent illuminated, waxing crescent
- Andromedids (**Dec 3**), ZHR=20, Moon 5 percent illuminated, waxing crescent

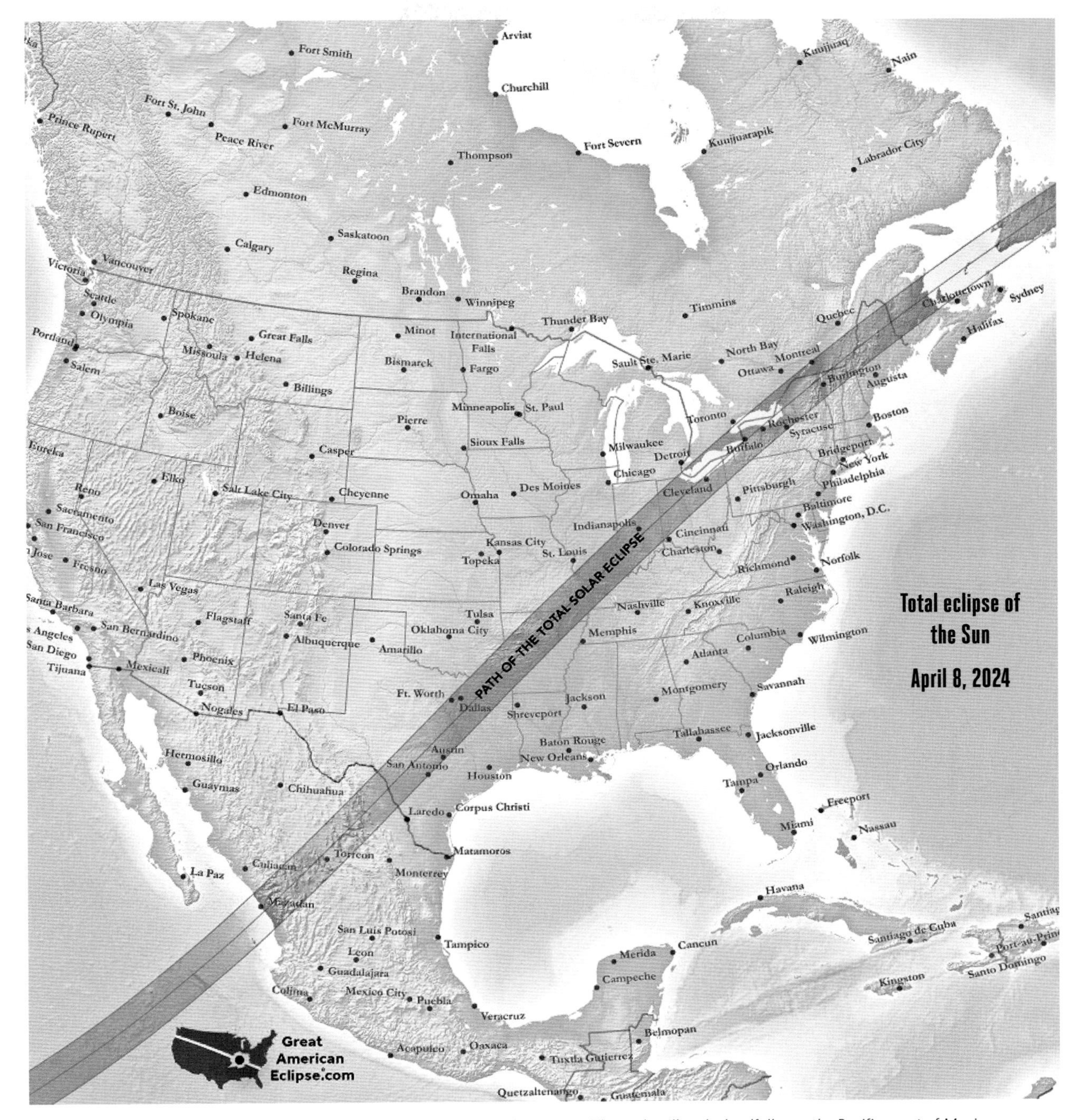

The path of totality for the April 8, 2024, total solar eclipse across North America. The path will make landfall over the Pacific coast of Mexico, cross into Texas and the central United States, then head out over New England and the Canadian Maritimes. Will someone—in the words of Carly Simon—take their "Learjet to (the north tip of) Nova Scotia, to see a total eclipse of the Sun?"

Comet C/2014 E2 Jacques makes a close pass near the Heart Nebula (IC 1805/NGC 896) in the constellation Cassiopeia on August 20, 2014.

THE OUTER PLANETS

The planet Mars does not reach opposition in 2024. Here are the opposition dates marking the best season to observe the remainder of the outer planets:

- Jupiter (**Dec 7**)
- Saturn (**Sep 8**)
- Uranus (**Nov 17**)
- Neptune (**Sep 21**)
- Pluto (**Jun 23**)

THE INNER PLANETS

The planet Venus does not reach greatest elongation in 2024. Mercury reaches greatest elongation seven times in 2024:

- Mercury (**Jan 12**) 24 degrees west (dawn)
- Mercury (**Mar 24**) 19 degrees east (dusk)
- Mercury (**May 9**) 26 degrees west (dawn)
- Mercury (**Jul 22**) 27 degrees east (dusk)
- Mercury (**Sep 5**) 18 degrees west (dawn)
- Mercury (**Nov 16**) 23 degrees east (dusk)
- Mercury (**Dec 25**) 22 degrees west (dawn)

ECLIPSES

There are four eclipses in 2024, two solar and two lunar:

- A penumbral lunar eclipse (**Mar 25**), maximum immersion of the Moon in the Earth's penumbra is 87 percent, favoring the Americas.
- A total solar eclipse (**Apr 8**), maximum duration of 4 minutes, 28 seconds, favoring North America.
- A partial lunar eclipse (**Sep 18**), maximum immersion of the Moon in the Earth's umbra is 9 percent, favoring the Americas, Europe, and Africa.
- An annular solar eclipse (**Oct 2**), maximum annularity of 7 minutes, 25 seconds, favoring Chile and southernmost Argentina.

CLOSEST CONJUNCTIONS

- Planet-planet: Mercury-Jupiter (**Jun 4**), 7 arcminutes apart, 12 degrees west of the Sun in the dawn.
- Planet-star: Mercury-Regulus (**Sep 9**), 30 arcminutes apart, 17 degrees west of the Sun in the dawn.

OCCULTATIONS

The Moon occults four planets fifteen times in 2024:

- Mercury (1): (**Mar 11**), South Pacific, +3 percent Moon
- Venus (2): (**Apr 7**), eastern North America (daytime), +2 percent Moon; (**Sep 5**), Antarctica, -6 percent Moon
- Mars (2): (**May 5**), Madagascar, -8 percent Moon; (**Dec 18**), Arctic, 87 percent Moon
- Saturn (10): (**Apr 6**), Antarctica, -7 percent Moon; (**May 3**), southern Indian Ocean, -26 percent Moon; (**May 31**), southernmost South America, -39 percent Moon; (**Jun 27**), northern New Zealand, -64 percent Moon; (**Jul 24**), Southeast Asia, -86 percent Moon; (**Aug 21**), northern South America, northwest Africa, -95 percent Moon; (**Sep 17**), western North America, +99 percent Moon; (**Oct 14**), India, eastern Africa, +89 percent Moon; (**Nov 11**), Central America, +77 percent Moon; (**Dec 8**), Japan, western Pacific, +51 percent Moon
- Best occultation (star): Spica (**Jul 14**) for North America. The Moon occults two bright stars twenty-two times in 2024: Spica (8) and Antares (14).
- Spica (8): (**Jun 16**), Russia, +73 percent Moon; (**Jul 14**), North America, +58 percent Moon; (**Aug 10**), Southeast Asia, +32 percent Moon; (**Sep 6**), eastern Africa, +12 percent Moon; (**Oct 3**), Hawaii (daytime), +1 percent Moon; (**Oct 31**), Eastern Europe, -1 percent Moon; (**Nov 27**), North America, -12 percent Moon; (**Dec 24**), Southeast Asia, -26 percent Moon
- Antares (14): (**Jan 8**), western North America, -11 percent Moon; (**Feb 5**), southwest Asia, -24 percent Moon; (**Mar 3**), Central America, Caribbean, -51 percent Moon; (**Mar 30**), Central Pacific, -77 percent Moon; (**Apr 26**), Saudi Arabia, east Africa, -93 percent Moon; (**May 24**), northern South America, Caribbean, -99 percent Moon; (**Jun 20**), Central Pacific, +98 percent Moon; (**Jul 17**), South Africa, +84 percent Moon; (**Aug 14**), South Pacific, +70 percent Moon; (**Sep 10**), western Australia, +45 percent Moon; (**Oct 7**), South Atlantic, +21 percent Moon; (**Nov 4**), southeast Pacific, +10 percent Moon; (**Dec 1**), South Africa, +1 percent Moon; (**Dec 28**), Central Pacific, -6 percent Moon

PERIODIC COMETS OVER MAGNITUDE +10

One periodic comet is predicted to break +10th magnitude in 2024: 12P/Pons-Brooks (perihelion **April 21**, magnitude +9) in the constellation Taurus

RESOURCES

The following references were used for compiling data for astronomical events for the next six years.

The American Meteor Society's Annual Meteor Showers: https://www.amsmeteors.org/meteor-showers/major-meteor-showers/

The British Astronomical Association Comet Section's list of future comets: https://www.ast.cam.ac.uk/~jds/

Greatest Elongations for Mercury and Venus: http://www.jgiesen.de/skymap/MercuryVenus/

Heavens-Above's Table of Planetary oppositions: www.heavens-above.com

NASA/GFSC Eclipse Web page: https://eclipse.gsfc.nasa.gov/eclipse.html

Occult version 4.2.0 software for Future Lunar Occultations of bright stars and planets: http://www.lunar-occultations.com/iota/occult4.htm

Stellarium: http://stellarium.org/

Seichii Yoshida's Weekly Information about Bright Comets: http://aerith.net/comet/weekly/current.html

MOON PHASES FOR 2024 (IN UNIVERSAL TIME)			
Closest Perigee: Mar 10 (356,893 km)		Most-Distant Apogee: Oct 2 (406,516 km)	
New Moon	Brown Lunation Number	Full Moon	Notes
Jan 11 - 11:58 UT	1250	Jan 25 - 17:55 UT	
Feb 9 - 23:01 UT	1251	Feb 24 - 12:31 UT	Minimoon
Mar 10 - 9:03 UT	1252	Mar 25 - 7:02 UT	
Apr 8 - 18:23 UT	1253	Apr 23 - 23:51 UT	Eclipse season
May 8 - 3:24 UT	1254	May 23 - 13:56 UT	
Jun 6 - 12:40 UT	1255	Jun 22 - 1:11 UT	
Jul 5 - 22:59 UT	1256	Jul 21 - 10:20 UT	
Aug 4 - 11:14 UT	1257	Aug 19 - 18:29 UT	
Sep 3 - 1:57 UT	1258	Sep 18 - 2:37 UT	
Oct 2 - 18:51 UT	1259	Oct 17 - 11:28 UT	Supermoon/Eclipse season
Nov 1 - 12:48 UT	1260	Nov 15 - 21:30 UT	
Dec 1 - 6:22 UT	1261	Dec 15 - 9:03 UT	
Dec 30 - 22:28 UT	1262 - Black Moon		

ACTIVITY: ASTRONOMY JOURNALING

Perhaps you're happy just checking out the sky tonight and knowing a little bit of what you're actually seeing. Or maybe you're perfecting your technique of photographing the night sky, or carefully counting meteors or measuring the brightness of a variable star, looking to contribute to some of the in-the-trenches grunt work of astronomy . . .

Keeping a journal is a great way to chronicle your path through astronomy. This can be on paper, electronic, or (I advise) a hybrid of both. Blogging or using social media to chronicle your observations are a great way to share what you're doing with the astronomical community and get immediate feedback. I like to note what I'm doing and observing on the social media platform Twitter each night, and share some of my latest astrophotography photos on Instagram. Twitter even lets you download a .txt archive of every tweet you've ever sent out, a handy searchable document I've utilized to retrieve information on just what I'm doing way back when on more than a few occasions.

Of course, keeping an exclusively electronic journal always runs the risk of falling prey to data death . . . could you lay your hands on a 3½-inch (9-cm) disk reader today, if you needed to?

A journal can be solely technical, or it can contain weather, cloud cover, sketches, thoughts on your place in time and space, and so on. I like to include time, date, location, a synopsis of what I'm observing, and perhaps a brief pencil sketch of our target for the night. Back in the days of film, it was standard for a photographer to carry a notebook and record the ISO, shutter speed, and f/stop for every shot, information that's now often simply embedded in the data file for each digital photograph.

I also designed a blank journal page some years back for easy data entry, along with blank fill-in disks for planetary sketches or eyepiece field of view(s).

Remember, you are working under the glow of a low-light red headlamp to fill in a journal in the field and sketch what you see; this is mainly why I like to keep entries and sketches as simple black-and-white drawings.

I can also see the progression curve in observational skill, looking far back over the decades. I remember some of those earliest entries, hastily sketching the shadow of the Earth on the Moon via projection during the total lunar eclipse on December 30, 1982. It was a typical frigid, northern Maine December morning, and I alternated between sketching the eclipse at the telescope and warming up inside by the wood stove.

Will anyone, centuries from now, come across my personal astronomy log and give it historical consideration? Possibly, though probably not. The same could be said for the deluge of ephemeral Tweets, Snapchats, and Facebook postings that fly around the world everyday today. Still, imagine the personal insights biographers might now have if, say, we had Einstein's or Isaac Newton's personal tweets.

You just never know. What's fleeting today could become crucial information tomorrow. I journal as much for myself as for the world, to recognize and acknowledge that I too once shared and pondered this brief corner of time and space.

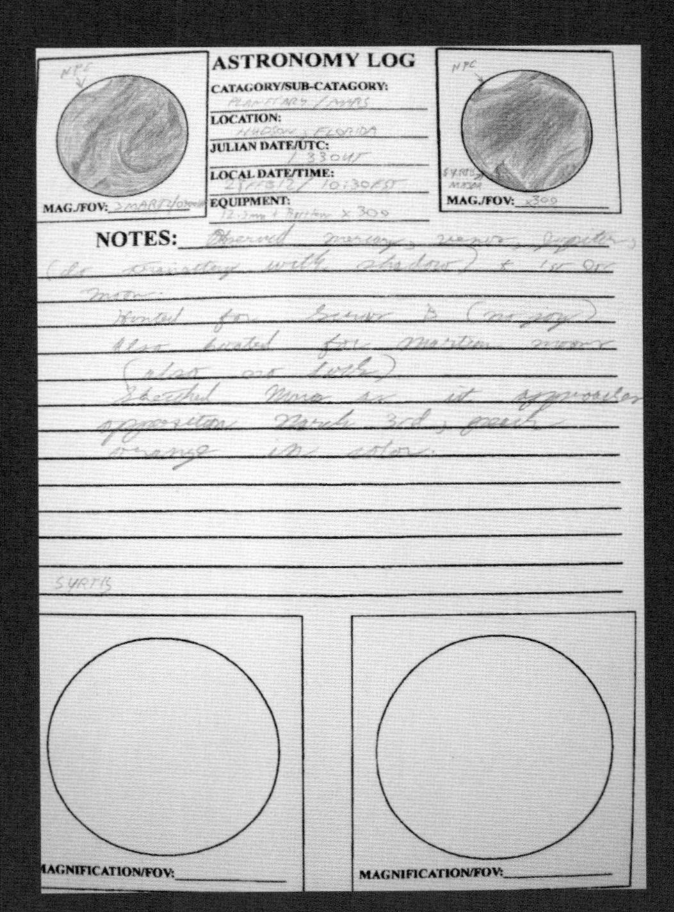

ASTRONOMY LOG

CATAGORY/SUB-CATAGORY:

LOCATION:

JULIAN DATE/UTC:

LOCAL DATE/TIME:

EQUIPMENT:

MAG./FOV:

MAG./FOV:

NOTES:

MAGNIFICATION/FOV:

MAGNIFICATION/FOV:

Our journal entry from an observation during the 2012 Mars opposition season. Having a blank template is handy for recording key data in the field . . . feel free to customize your own. I like having two disk templates, one smaller pair (top) for planetary disk sketching, and a larger set (bottom) for deep sky/star field sketching.

ACTIVITY: RUNNING AND JOINING A STAR PARTY

Looking to network with other amateur astronomers? Got a new telescope for Christmas and haven't a clue as to how to use it?

One of the best things I've ever done was join a star party. Not only does this allow me to check out new gear and gadgets, but it offers me a chance to see what neighboring astros are up to, be it with deep sky imaging projects, or newly built telescope rigs. And heck, it keeps me on my toes and on top of the very latest in space news and what's up in the sky.

Plus, amateur astronomers just love to talk astronomy. I'll admit, while it's fun to run a telescope at such an event, I also have fun just walking around in the dark and checking things out *sans* telescope to anchor me down. I always search out the local astronomy club while I'm traveling to see what's up. This habit came in handy, for example, the first time I traveled to Australia without a telescope. The local club near Ayers Rock was more than happy to show me around the southern hemisphere sky.

Attending

Here are some dos and don'ts that telescope operators want the general public to know:

- Red lights aren't mandatory at public star parties, but they're preferred. Likewise, arrive before dark, to avoid blinding everyone with your car's high-beams.
- Apply bug spray (or sunblock, during a solar viewing event) away from a telescope's optics.
- Everyone's eyes and visual acuity are slightly different, and (most) telescope operators know this and are skilled at guiding folks to the eyepiece and coaching them through looking through it. If you normally wear glasses, use them while looking through a telescope. Remember that you're allowed to tweak the focus slightly until the image is sharp. If you see nothing, say so, and the telescope operator will quickly line the target back up.
- Ask us anything. I've had spirited discussions in the dark with folks about aliens, astrology, star-naming, the latest space news, the current end-of-the-world-of-the-week, you name it.

The next logical step is bringing your own telescope to the party and showing off the cosmos. Most parties are pretty laid back affairs . . . a few will actually organize and assign scope operators different objects. This is handy, and assures that A) you can read up on your assigned object beforehand and be knowledgeable, and B) you won't hear someone say, "Oh yeah, the Orion Nebula. We saw that in the last five telescopes." The downside is it gets *boring* showing off M13 all night. We generally can tell what everyone else is looking at by looking down the row of telescopes pointed skyward, and seeing what all of the telescope tubes are aimed at. Then we aim at something of interest in the opposite direction, and give 'em something new.

Is there a good International Space Station pass tonight? How about an Iridium flare, or a binocular comet? We always print out the satellite passes for the night from Heavens-Above, and make a note of any other exceptional objects. We once surprised a grade school group in Florida with a nighttime satellite launch off the Space Coast that just happened to coincide with the star party!

Never give up. We've chased holes in the clouds and shown folks targets of opportunity. We've done successful sidewalk star parties from light-polluted downtown Tampa and St Pete. We've also experienced the quixotic defeat of joining a group of amateur astronomers as we sat in our cars in a row on an otherwise empty field, waiting in vain for the rain to pass. If possible, plan a rainy night presentation indoors if needed.

Don't miss a chance to join a star party and experience the wonders of the cosmos. I might just show up at YOUR local star party next.

The author with the 23-inch (58-cm) "War of the Worlds" refractor at the Charles E. Daniel Observatory at the Roper Mountain Science Center in Greenville, South Carolina.

(Right) Humans, contemplating the Universe . . . the Milky Way shines over Petrova Gora in 2012, during the most popular annual star party in Croatia.

REAL SCIENCE YOU CAN DO AND PROTECTING THE NIGHT SKY

Doing science from your own backyard.

"All truths are easy to understand, once you discover them. The point is to discover them."

— Galileo Galilei

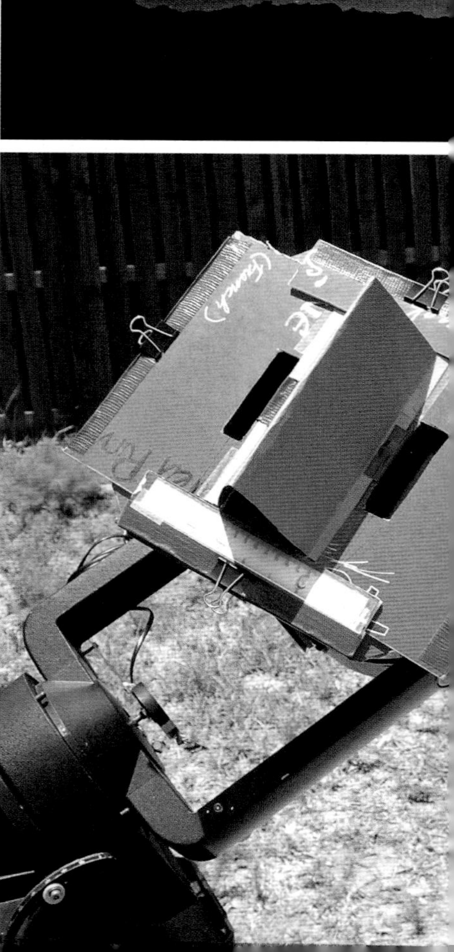

ACTIVITIES

- How dark is my sky? How to gauge the quality of the night sky (page 208)
- Building a cardboard interferometer (page 210)
- Renting and using a telescope online (page 211)

TIME ON PROFESSIONAL TELESCOPES IS PRECIOUS AND SOUGHT-AFTER

The Andromeda Galaxy (Messier 31) 2.5 million light-years distant. Composed using 40x60-second frames, focal length 388mm f/6.3, ISO 3200. Image credit and copyright: Dan Sullivan. Instagram: @dansullivan.

Left corner: Observing meteor showers and reporting what you see is still a great way for amatuer astronomers to contribute to real science. Pictured: a composite of the 2015 Geminids over Mount Kanin in northwestern Slovenia.

Modern professional astronomy is a relative newcomer on the scientific scene. Before the late eighteenth century, science in general and astronomy in particular was the sole domain of hobbyists, dabblers with the time and—very occasionally, such as in the case of Sir William Herschel—the grace of a patron such as King George III to fund their scientific endeavors.

The interplay of amateur and professional astronomer (sometimes simply called a "pro/am collaboration") is alive and well today. Time on professional telescopes is precious and sought-after, and some amateur astronomers now have access to large telescopes in backyard observatories that would be the envy of many universities. Many advanced, American astronomy enthusiasts bristle at the label of *amateur*, as it's often equated with novice or neophyte (as in the term a "rank amateur"). Throughout this book, we use the term "amateur" more in the British sense, as one who does something for the pure love of it. From observing variable stars to discovering comets, there are lots of opportunities for amateurs to make real contributions to astronomy.

Thus far, we've touched on some of the things amateur astronomers still make scientific contributions to, including the following:

Discovering and monitoring comets: Sure, there's lots of competition out there these days from automated sky surveys and other robotic eyes scouting the sky, but backyard observers can and do still occasionally nab a comet. (See Chapter 7, How to Report a [Potential] Comet Discovery, page 128.) Southern hemisphere observers still have a distinct advantage, as most of the surveys hunting for Near Earth Objects (NEOs) are northern hemisphere-based. Earth-based surveys still have sunward blind spots, a fact made painfully clear on the morning of February 15, 2013, when a meteor exploded over the city of Chelyabinsk, Russia, without warning. About 1,500 people received minor injuries (mostly from flying glass),[1] but thankfully, no one was killed.

Sometimes amateurs just get lucky and happen to notice the fuzzball of a new comet during routine imaging or observing. Dedicated sweeps for comets usually target the ecliptic in predawn or post-dusk hours—likely hiding grounds for a new comet—though these days, it's also a worthy pursuit to widen the search to high-inclination areas or dense star fields, regions where automated surveys may still miss an icy interloper.

Counting meteors: As mentioned in Chapter 7 (Observing Meteor Showers, page 130), you can simply sit out and count how many meteors you see from a given meteor shower. Meteoritics (the study of meteors and meteor showers) is still working on understanding and modeling meteor shower streams. If an ancient shower stream has yet to make its presence known by intersecting the path of the Earth—even if it's part of a known annual shower—then astronomers will have little clue that it exists. Ninety-five annual showers—most of which are faint, with Zenithal Hourly Rates (ZHR) of less than five per hour—are known or recognized by the International Astronomical Union,[2] though new ones are occasionally identified. Streams evolve over centuries, with older ones fading out and new ones developing. If you notice an uptick in meteors coming from the same direction or a new radiant in the sky on a given night, you might just be witness to the birth of a new and undiscovered meteor shower.

Variable star observing: Amateurs have long contributed to the science of variable star observing. Arab astronomers referred to the notorious naked eye variable star Beta Persei in the constellation of Perseus as "Algol" or the Demon Star, perhaps a reference to the fading act it pulls every 2.9 days. Today, we know that Algol is an **eclipsing binary**, with one star passing in front of the other along our line of sight, one of thousands of such variable stars known. Today, pro/am astronomers monitor everything from cataclysmic variables to novae and pulsating variable stars. Variable stars not only paint a coherent picture of modern, descriptive astrophysics and the lives of stars, but they also serve as cosmic yardsticks: In 1912, the astronomer Henrietta Swan Levitt noted a particular class of variable star, known as a **Cepheid variable**,[3] has a direct relationship in brightness versus pulsation period. Find a Cepheid in a galaxy, and you can use the inverse square law (that is, light drops off, say, by ¼ at twice the distance, ⅑ at three times the distance, etc.) as a useful cosmic yardstick.

Amateur astronomers are also often the first to raise the alarm when a new nova appears in our galaxy. A good naked eye nova occurs, on average, about once a decade, though binocular novae brighter than about +10th magnitude occur a few times every year, mostly along the densely populated galactic plane.

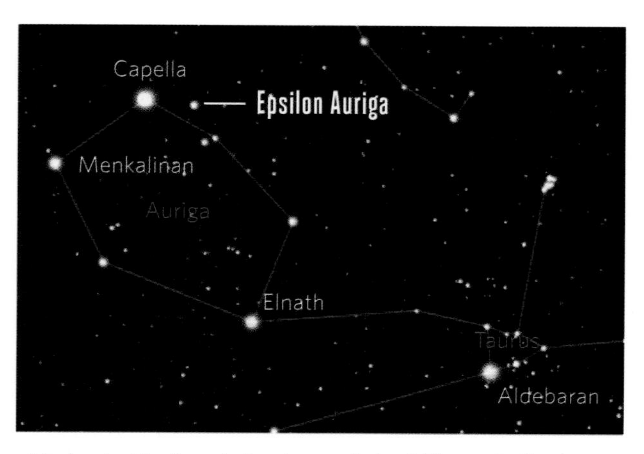

A finder chart for the notorious long-period variable star, Epsilon Auriga. Located near the bright star Capella (pictured) in the constellation Auriga the Charioteer.

Sporadic **recurrent novae** such as T Pyxidis are also fascinating to keep tabs on. On the night of January 28, 2010, the Florida-based amateur astronomer Barbara Harris raised the alert that recurrent nova U Scorpii was once again active.[4]

Astronomers at Calvin College also predict that the +12th magnitude star KIC 9832227 in the constellation Cygnus the Swan may go nova in 2022 on or thereabouts,[5] something worth keeping an eye out for. The familiar cruciform shape of Cygnus will appear deformed if or when it does!

Long before the mystery of Tabby's Star (KIC 8462852), amateur astronomers also kept tabs on the enigmatic variability of Epsilon Aurigae in the constellation of Auriga the Charioteer. This +2.9 magnitude naked eye star occasionally dips down a full magnitude once every 27 years . . . and stays there for an amazing 640 to 730 days. The best theory: Epsilon Aurigae has a small stellar companion, enshrouded in a dense debris disk. The next dip in brightness for this fascinating star is expected to occur in 2037.

The main clearinghouse for variable star observations is the AAVSO (the American Association of Variable Star Observers) which has logged an amazing twenty million observations plus since its founding in 1911.

You can make variable star observations simply by noting the brightness of the target star versus the brightness of known stars. (The AAVSO provides printable finder charts with comparison stars.) You can also employ a DSLR camera as a ready-made imager for variable star photometry using the AAVSO's guide: https://www.aavso.org/dslr-observing-manual

And while we think of deep sky objects as unchanging, things do vary on timescales you might expect to live through, especially when it comes to reflection nebulae illuminated by internal stars. NGC 1555 (Hind's Variable Nebula), Gum 1-29 (Gyulbudaghian's Nebula), and Hubble's variable nebula (NGC 2261) are all examples of nebulae which occasionally change in brightness.

In 2004, Jay McNeil noticed a new brightening nebulosity in the corner of the Messier 78 nebula in the constellation of Orion the Hunter, a nebula which now bears his name as McNeil's Nebula.[6]

Planetary impact alerts: You would think, by now, there wasn't much left to discover in our very own backyard of the Solar System, but seasoned amateur observers do occasionally spot real changes on planetary surfaces and atmospheres. One famous, recent example is the spokes in the rings of Saturn spotted by the amateur astronomer Stephen O'Meara in the 1970s, and subsequently confirmed by the Voyager 1 mission during its 1980 flyby.

Satellite tracking: Whenever a new classified satellite is launched, backyard satellite trackers are on the hunt, looking to characterize its orbit and confirm or deny its existence. For example, Russia launched a mission known only as Kosmos-2499 in 2014, a satellite which made repeated erratic maneuvers and was thought to be a clandestine test of either an on-orbit refuel, repair, or perhaps anti-satellite killer technology. Veteran satellite-spotter Marco Langbroek also chronicled the recent approach of USA-276 near the International Space Station, thought to be testing similar technologies. These are all visible from your backyard, if you know when and where to look. Heavens-Above, Celestrak, and the SeeSat-L message board are great resources to aid in your quest.

Tracking and recording phenomena during total solar eclipses and occultations: The simple act of one thing passing in front of another actually conveys lots of information to astronomers. During the August 21, 2017, total solar eclipse, thousands of observers participated in the Eclipse Megamovie and Citizen CATE projects in an effort to understand the elusive solar corona. Likewise, occultations of bright stars by the Moon (see Chapter 3, Occultations, page 66) can not only help to map the profile of the lunar limb, but also reveal the existence of close double stars. And if enough observers witness the occultation of a star by an asteroid, a shape profile emerges, perhaps revealing a binary nature or undiscovered ring or moonlet.

For example, observations of the remote Centaur asteroid 10199 Chariklo in 2014 revealed it had a ring. Another example is the amateur campaign to observe Kuiper Belt Object 2014 MU69 ahead of New Horizon's New Year's Day 2019 flyby. This campaign revealed the possible binary contact nature of 2014 MU69.

Steve Preston maintains a list of upcoming asteroid occultations online. Keep in mind, there's still a small bit of uncertainty involved with these events, even today. Even though you're observing from along the predicted path, you still might not see anything at all! Be sure to familiarize yourself with the field of view prior to the event, and make sure you've got the right target star selected. We like to have the WWV radio broadcast out of Fort Collins, Colorado, running in the background, for a precise time hack leading up to the event. Arguably the most boring radio station in the world, WWV simply announces the time at the beginning of each minute, and clicks for each second. WWV radio has broadcasted the time since 1920 on AM shortwave frequencies 5, 10, 15, and 20 MH_2.

What's left to learn from occultations? Well, in the case of the Moon, a map of the jagged lunar profile can be assembled, using numerous observations of grazing passes of a bright star. These events can be especially dramatic, as the star seems to wink in and out from behind lunar peaks as it shines down lunar valleys along the observer's line of sight, proof that the jagged Moon isn't exactly round. This is similar to the effect known as Baily's Beads, which occurs during a total solar eclipse.

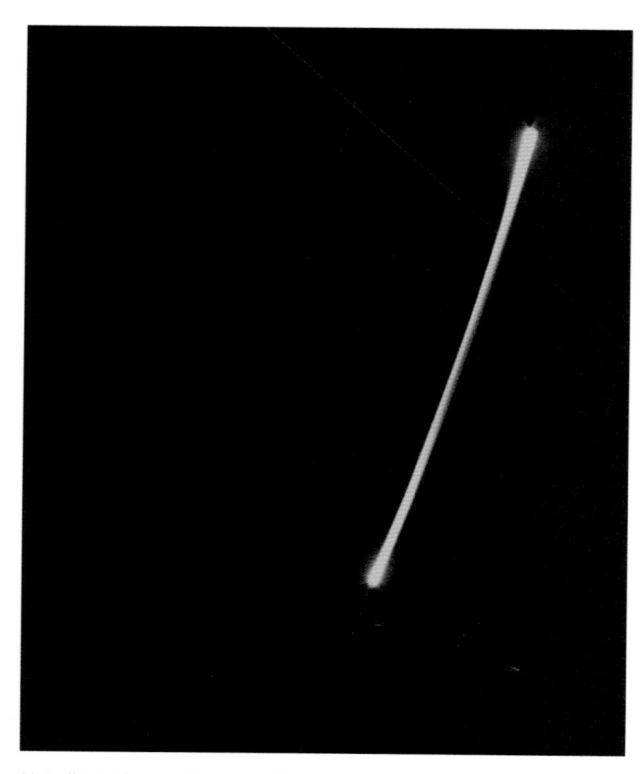

Not all total lunar eclipses are the same. This single long exposure shows the change in the color of the Moon through all the stages of a lunar eclipse. The color of the eclipsed Moon during totality can be described by its Danjon Number, with 0 as very dark brick red, through 4 as a yellowy, bright eclipse. The color and brightness of the eclipsed Moon are both a function of the level of dust and aerosols in the Earth's atmosphere at the time, something amateurs can monitor and note during an eclipse. One single film exposure at high f/stop, digitally scanned.

Extremely close double stars can also reveal themselves during a lunar occultation as well, as one winks out behind the Moon just before the other. Incidentally, Regulus has long been suspected of harboring a +13th magnitude white dwarf companion, which might just briefly make an appearance during a lunar occultation. Likewise for the mysterious and often reported **Ashen Light of Venus**, which could—if truly real and not merely an optical phenomenon—be briefly visible during a favorable occultation by the Moon against a relatively dark sky.

In the case of asteroid occultations, measurements from enough locations can create chords showing the shadow profile of the occluding space rock. If a series of observation chords are collected, a silhouette model of the asteroid can be made, measuring the size and shape of the asteroid and perhaps even revealing the presence of any tiny moonlet companions. For example, in 2017 an amateur effort to observe asteroid 113 Amalthea revealed that it very probably has a tiny moon.

For an asteroid occultation, you'll need a camera that can go much fainter than most off the shelf models. Many occultation chasers now modify low-light security cameras for this effort, though more expensive MallinCams and other specialty video cameras are designed specially for low-light level astronomy.

The International Occultation Timing Association (IOTA) always welcomes observations of occultations. The effort also lends itself to video astronomy, and while you can make visual observations like those astronomers of old, video observations with precise time stamps and the GPS coordinates of the observer's site are preferred. For lunar occultations, the effort is as simple as aiming a video camera afocally through the eyepiece just before the appointed time. Your biggest challenge will be to balance the image exposure between the dazzling daytime side of the Moon and the relatively dim star or planet.

Amateur astronomers capable of doing photometric measurements can also perform backup observations of exoplanet candidates via a campaign known as TransitSearch.org: http://www.transitsearch.org/

ENTER ONLINE ASTRONOMY

Got clouds? The very first time I saw someone running the SETI at Home screensaver on their computer back in the late 1990s, I realized I just *had* to get a PC. In the past decade, astronomy has moved online in a big way, with volunteers doing everything from classifying galaxies to hunting for exoplanets and mapping craters.

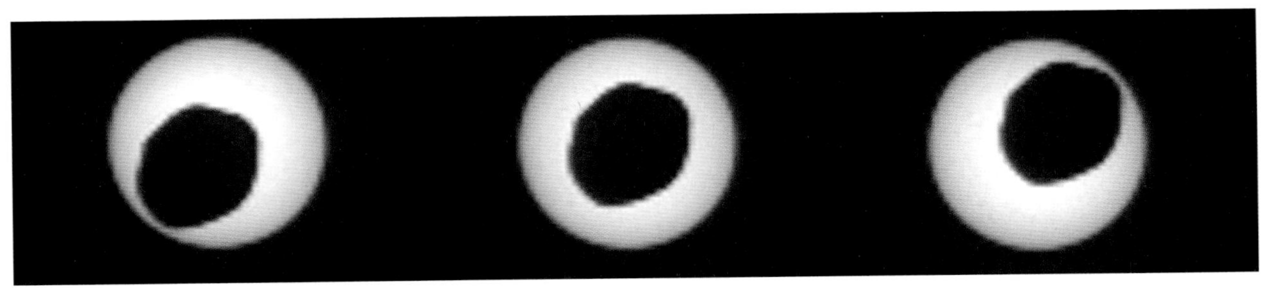

Robots on other planets do some surreptitious astronomy on occasion, too . . . such as this sequence showing a curious "potato-shaped" annular solar eclipse as the Martian moon Phobos crosses in front of the Sun, captured by the Curiosity rover on August 20, 2013. What other amazing scenes await the patient armchair sleuth in the flood of spacecraft images available in the public domain?

GALAXY ZOO AND THE ZOONIVERSE

It started with a simple idea over a few beers at the Royal Oak Pub in Oxford, England. The astronomers Chris Lintott and Kevin Schawinski came up with an innovative solution to a key problem: how to crowdsource and automate the tedious process of classifying the hundreds of thousands of galaxies in the Sloan Digital Sky Survey. And thus Galaxy Zoo—the first in an avalanche of citizen science projects—was born. After a brief tutorial, anyone with a laptop and a web connection could view images and describe what they saw in those distant galaxies. Galaxy Zoo and its ilk work because they rely on a preponderance or consensus of data; in other words, one viewer might erroneously catalog a galaxy as a spiral, while hundreds of other volunteers all agree that it's an elliptical.

But what was really intriguing was how Galaxy Zoo volunteers found the unexpected, in the truest spirit of science. For example, the Dutch schoolteacher-turned-volunteer Hanny van Arkel noticed a strange green patch near a galaxy in an image, something a robotic eye might've ignored. Discussion of Hanny's Voorwerp (Dutch for "Hanny's Object") actually caught the eye of professional astronomers enough to merit observation time using the Hubble Space Telescope. Astronomers later realized that this seemingly insignificant blob is an entirely new class of object, a quasar ionization echo.

Today, the Galaxy Zoo Zooniverse contains a menagerie of crowdsourced projects, including the original Galaxy Zoo, Solar Storm Watch, and the Milky Way Project, which looks for bubbles in the interstellar medium.

My favorite is Old Weather, which includes ship's logs from around the world. It was standard procedure for ships at sea to record weather data, and these logs go back over 150 years. Occasionally, a bit of undiscovered astronomy crept into these logs as well, awaiting rediscovery.

Join the Zooniverse: https://www.zooniverse.org/

Also, check out Cosmoquest for more great Citizen Science programs, including Moon Mappers, Image Detective, and more: https://cosmoquest.org/x/

PLANETARY IMAGE PROCESSING

All of the raw data imagery from ongoing missions, such as Curiosity on Mars or Juno in orbit around Jupiter, are readily available for anyone wanting to take the time and sift through them. But beyond fanciful pareidolia claims of mini-Bigfoots or flowers on Mars, the rovers or orbiters on or around far-flung worlds see some fascinating things.

One great example is the occasional misshapen annular eclipses involving the tiny moons Phobos and Deimos witnessed by rovers on Mars.

NASA even encouraged amateur image processors to cull the best of the best during its Juno mission before the orbiter's demise in July 2018. Follow Jason Major (@JPMajor) and Kevin Gill (@kevinmgill) to see what's possible.

There are also real discoveries in NASA images awaiting the diligent. Sometimes, these discoveries are made in images years or even decades old. For example, the amateur image processor Ted Stryk noticed images of a shadow transit of Neptune's moon Despina sitting in Voyager 2 archives . . . 20 years after the flyby and discovery of the tiny moon. Certainly, such a discovery is possible; in fact, one curious aspect of the New Horizons July 2015 flyby of the Pluto-Charon system was that it discovered *no* new moons at all. Did we miss something?

THE RISE OF AUTOMATED SURVEYS

Get ready: astronomy is set to grow online over the next decade in a big way. New surveys are on the rise, promising to flood professional astronomers with a deluge of data. The LSST, for example, will see first light in 2019, and will survey the entire sky visible from its vantage point high in the Andes Mountains down to +27th magnitude and generate an expected 1.28 petabytes worth of data, every year. That's lots of faint comets, asteroids, and variable stars just waiting to be found. The current ongoing Gaia survey placed in space by the European Space Agency wrapped up its primary astrometric mission in 2018, mapping the positions of millions of stars with unparalleled precision. The PanSTARRS 1 survey—already a prolific comet hunter—will give way to the larger PanSTARRS 4 (as in four telescopes) with more exciting discoveries and data to sift through to come.

We're about to enter the true golden age of online astronomy over the next decade.

GET READY:

astronomy is set to grow online over the next decade in a big way.

SOCIAL MEDIA

Don't forget social media's role in amateur astronomy. Today, when a bright reentry or a Chelyabinsk-style event occurs, we're hearing about it first on social media platforms such as Twitter. It's always encouraging to see our Twitter stream flooded with images of the latest eclipse or comet, crowding out the endless background noise of political and entertainment chatter . . . hey, maybe there's hope for humanity yet!

Read more about online pro/am collaborations courtesy of the astronomer Pamela Gay:

http://www.skyandtelescope.com/get-involved/pro-am-collaboration/how-to-get-involved-in-amateur-research/

LIGHT POLLUTION:

THE SHINY ENEMY

It's a disheartening battle. Growing up in rural northern Maine, I was fortunate to have dark skies right in my own backyard. Summer meant camping out on Eagle Lake where skies were darker still, with a star-studded Milky Way extending from the horizon to the zenith.

Today, such skies are a tough find in the developed world, especially in Europe, Northeastern Asia, and the overlit U.S. Eastern Seaboard. The culprit is **light pollution**, and the drive in developing countries always seems to trend toward abolishing the night. It's even been a proposed method for detecting alien civilizations: Simply look for the lights of their cities, dappling the nighttime side of their homeworld.[7]

A discouraging nighttime view: light pollution over the contiguous United States as seen from 2012. This is a nighttime summation/composite view, which is why there are no clouds in the image. The message: go west and away from the bright urban lights, dear hopeful astronomer.

What's really sad is that no one seems to miss those dark, star-filled skies. Modern lore is dotted with tales of urban dwellers unfamiliar with the sky, calling in the Milky Way as smoke on the horizon during a blackout.

We've lost familiarity with nature, for sure. But enough of the depressing part. What can you do about it?

Sky watchers have always had some natural light pollution to contend with throughout history, in the form of the Sun and the Moon. Modern urban lighting has done more than its fair share of disruption with this natural cycle. The International Dark Sky Association (IDA) remains committed to raising awareness concerning light pollution since it was founded in 1988.

WHAT YOU CAN DO

Most non-stargazing people tend to roll their eyes at the mere mention of the term "light pollution." In a world where we're told everything is potentially hazardous, here's one more thing on the long list of modern worries and fears. Many communities have ordinances on the books against **light trespass**, though these, like so many other codes and laws, are only enforced if someone takes the initiative to advocate for them. There aren't (yet) any light pollution police, though some communities of like-minded amateur astronomers have gotten together in places such as Sky Village near Portal, Arizona, and Chiefland, Florida, and formed dark sky villages.

We've often defused a potential neighborly light trespass dispute by simply inviting friends over and showing them the night sky.

PERILS OF LIGHT POLLUTION

And it turns out that light pollution doesn't just impact amateur astronomers. It also affects:

Health: The American Medical Association (AMA) announced in 2016 that light pollution and chronic sleep deprivation affects health, elevating the risk for diabetes, cancer, and cardiovascular disease.

Wildlife: Light pollution upsets the circadian rhythms of nocturnal animals as well, including migratory birds disoriented by overlit billboards, beacons, and buildings.[8]

The Economy: This key effect resonates with many, even individuals who otherwise would care little about preserving the night sky. Lots of wasted light goes towards illuminating the undersides of clouds and aircraft flying overhead. An estimated U.S. $7 billion is wasted on nighttime over-illumination every year. Local communities discover big savings by replacing garish overlit areas with smart lighting that only comes on for a brief few hours after dusk or just before dawn, when human activity is at its peak. For businesses and communities interested in the bottom line, this sole factor can play a big role.

GOOD NIGHTTIME LIGHTING CAN STILL BE SMART AND EFFECTIVE,

but dark sky friendly.

Safety: Personal safety and security are often cited as top factors for nighttime lighting, though there's good evidence to suggest that crime rates actually *increase* thanks to overlighting. Crooks need to see what they're doing to commit crimes.

National security: Finally, light pollution affects national security assets. For example, nearly every mountain peak surrounding Tucson, Arizona, has a major observatory complex on top of it. These assets are paid for by tax dollars, to include not only scientific resources that keep America on the cutting edge of scientific research, but also Department of Defense assets involved in tracking satellites. These rely on dark skies, and the expansion of the city of Tucson and its encroaching light pollution has had a direct impact on the continued viability of these sites.

GOOD FIXTURES VERSUS BAD FIXTURES

It doesn't have to be this way. Good nighttime lighting can still be smart and effective, but dark sky friendly. The reality is you can buy light fixtures at local hardware stores that contravene local light-pollution ordinances. A good dark sky–friendly light fixture aims the beam downward, toward what it's meant to illuminate. Preexisting lights can often also be easily and cheaply retrofitted with a hood or bezel, turning a former light-pollution offender into a dark sky–friendly light fixture.

Dueling bands of natural sky glow, the southern band of the Milky Way versus the zodiacal light.

ACTIVITY: HOW DARK IS MY SKY? HOW TO GAUGE THE QUALITY OF THE NIGHT SKY

How dark is the sky from your backyard . . . really? You may be surprised to know that you can easily gauge your limiting magnitude simply by counting stars. The darker the skies are, the more stars you can see.

There are some key regions of the northern hemisphere sky that are well-suited to performing a quick star count. To make this simple but effective exercise work, said target region should be transiting (i.e., highest in the sky) near the north-south meridian when you make the measurement.

One great gauge is the Square of Pegasus. The square asterism is within the boundaries of the extended astronomical constellation of Pegasus, the winged horse of Greek mythology. The corners are bounded by four +2.5 magnitude stars. If you can't even see these, you're under *very* bright skies, indeed!

Generally speaking, if you can count (A) number of stars in the Great Square (not to include the four corner stars), then your limiting magnitude is (B):

- (A) number of stars = limiting magnitude (B)
- 0 stars = limiting magnitude of 4 or brighter;
- 4 stars = limiting magnitude of 5;
- 14 stars = limiting magnitude of 6.

Pegasus not up in tonight's sky? Another good, familiar winter target is the constellation Orion. Within the confines of the square bounded by Betelgeuse in the upper-right shoulder of the Hunter (facing the observer) down to Rigel in the Hunter's knee is another good gauge box for estimating the limiting magnitude of the night sky. Including the three famous belt stars, can you count:

- 9 stars = limiting magnitude of 4;
- 11 stars = limiting magnitude of 5;
- 15 stars = limiting magnitude of 6.

The Pleiades (M45): The Open Cluster Messier 45 (see Chapter 5: The Sky From Season to Season, page 80) is also known as the Seven Sisters for a reason . . . can you count all seven? The faintest is Celaeno at magnitude +5.4. It may seem amazing to think, but a keen-eyed observer can actually spy sixteen stars in M45 under a dark +6th magnitude limited sky.

Finally, here's a unique and time-honored eye test: can you split the stars Alcor and Mizar in the bend of the handle of the famous Big Dipper asterism? Long before there was an eye test at the Department of Motor Vehicles, Arab astronomers referred to using this pairing as a sort of cosmic eye test. Can you split the pair? Alcor and Mizar are 0.2 degrees or 12 arcminutes apart, a pretty easy split.

You can delve in depth into Pegasus Square sky test at: https://freestarcharts.com/how-dark-are-your-night-skies

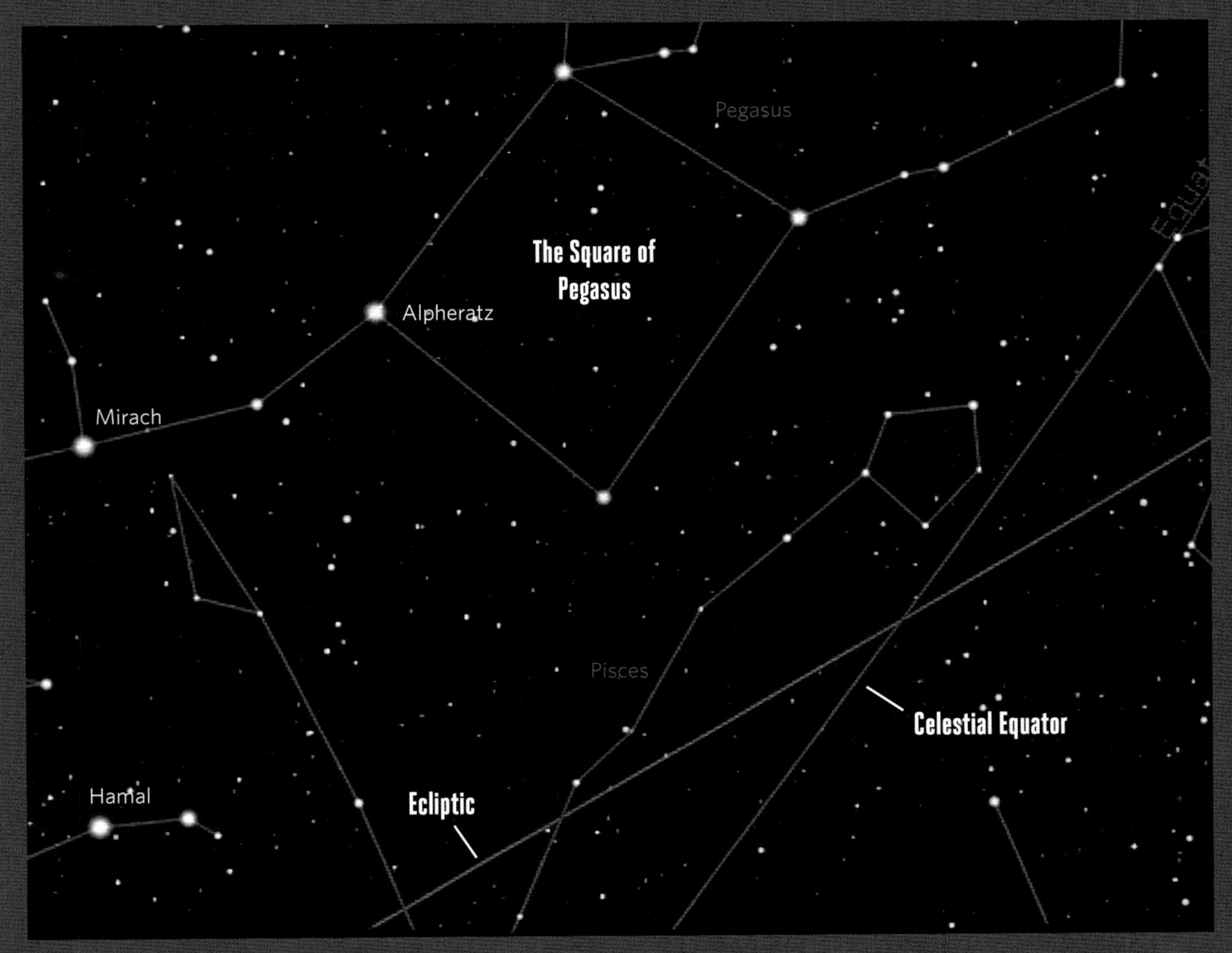

A finder chart for the Square of Pegasus asterism in the extended constellation of the same name as seen from latitude 30 degrees north on September 1 at 10 p.m. local, looking east.

ACTIVITY: BUILDING A CARDBOARD INTERFEROMETER

I'm about to introduce you to the strangest-looking modification you can make to your telescope.

Double stars are fascinating to observe over the span of several years, and along with variable stars, represent one of the few things that actually *change* in the deep sky over the span of a mere mortal lifetime.

The problem is, when it comes to close double stars, resolution and acuity usually runs up against our friend the Dawes Limit (see Chapter 2, page 32) at around under 1 arcsecond of separation.

There are two nifty tricks professional astronomers can use to untangle really close binary stars, known as **spectroscopy** and **interferometry**. You can actually use interferometry (and a little math) to measure the position angle and separation of really close binary stars.

Interferometry actually uses the dual particle/wave nature of light, contrasting and combining the two using a filter mask placed in front of the instrument. Two caveats: This method works primarily with two stars under 1-arcsecond separation, which are equal (within one magnitude) in brightness.

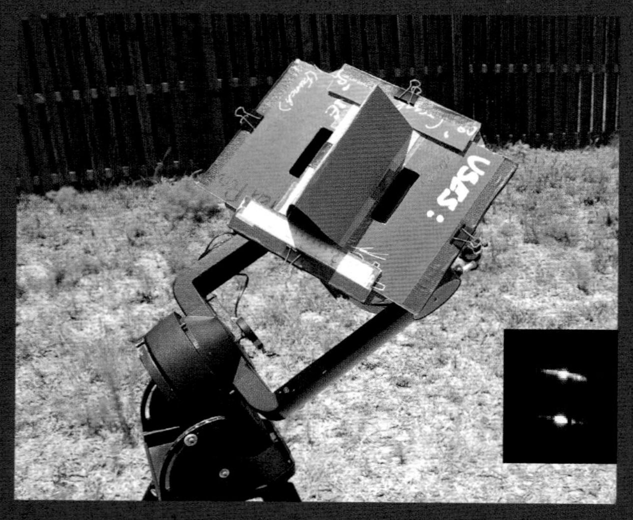

One of the most curious-looking telescope masks I've ever built: a cardboard interferometer. The inset shows what the interference fringes look like on the computer screen or through the eyepiece, from interfering (top) to non-interfering (bottom).

Making the Mask

The collar for the mask is similar to the one explained in the Making a Safe Homemade Solar Filter for Binoculars or a Telescope section in Chapter 8 (page 148). However, you want this collar to be a bit loose, as it needs to rotate by hand.

The second mask looks like, well, a mask, with two 30 mm by 90 mm interferometer slits cut in it. Be forewarned: these slits will also dim the target star substantially; I wouldn't make a cardboard interferometer for a telescope smaller than a 6-inch (15-cm) reflector.

The inner mask slides along tracks mounted on the edge of the collar. The key is knowing the spacing of the slits during the observations.

Using the Mask

Using the interferometer requires patience. What you're watching for are fringes, an interference strip of light seen at the eyepiece caused by the two stars.

Once you've acquired, centered, and focused the target star, put the mask on the front of the scope. Set the aperture to the narrowest spacing between the two slits. Then, rotate the mask until the fringes (stripes) seem brightest in contrast and begin to overlap. It helps if you can have a friend very carefully rotate the mask in small 30-degree increments while you look through the eyepiece. This measurement taken along the collar in degrees of arc is the **position angle** of the two intertwined stars.

Now, slowly vary the two slits in distance from each other, until they interfere, or the fringe stripes appear to overlap with each other as one continuous band of light. That's where you take the crucial slit spacing measurement.

Now to plug the spacing measurement into a meaningful observation.

Use the equation: $s = 56.72/d$, where 56.72 is a function of the wavelength of light, d is the distance between the two slits measured from their centers in millimeters, and s is the measured separation of the double stars in arcseconds.

Star	Coordinates (R.A./Dec.)	Magnitude(s)	Position Angle/ Separation	Orbit (in years)
Gamma Centauri	12H 42′ -48° 58′	+2.9	349°/1.1″	85
Eta Ophiuchi	17H 10′ -15° 43′	+3.2	239°/0.6″	88
Zeta Sagittarii	19H 3′ -29° 53′	+3.3	244°/0.6″	21
Gamma Lupi	15H 35′ -41° 10′	+3.5	274°/0.7″	190
Zeta Boötis	14H 41′ +13° 44′	+4.5	301°/0.9″	45
Zeta Cancri AB	8H 12′ +17° 39′	+5.6	104°/0.8″	60

Variations

I've seen some variations on the cardboard interferometer mask over the years. A few involve a grated mask with multiple slits, a plus in terms of combating light loss and dimming. I still like the slit approach, however, in terms of its simplicity, especially when working the final equation.

Follow a set of double stars in terms of position angle and separation for a few years, and an orbit emerges.

Here are some good close double stars of equal magnitude to practice on.

Note: While it's the faintest of the selected stars, Zeta Cancri provides an interesting contrast, as you can see the third solitary companion along with the close two primary stars in the same field of view. This third star gives you a handy gauge while looking at interfering versus non-interfering star fringes.

ACTIVITY: RENTING AND USING A TELESCOPE ONLINE

Clouded out? An increasingly viable option is to simply rent a telescope online. Most of these are automated telescopes based out of sites such as the Canary Islands off the coast of Africa in the Atlantic or in Mayhill, New Mexico, regions offering unparalleled clear skies 360 days out of the year.

The two leaders in the rent-a-scope market are Slooh (https://www.slooh.com/) and iTelescope (http://www.itelescope.net/). These both give the users the ability to remote control telescopes and choose targets, and offer free trial subscriptions. This also gives users a chance to practice using post-processing utilities such as Photoshop.

One thing to keep in mind: Many of these services retain the copyright for any images obtained using these instruments, usually not a problem for casual observing, but a situation that could get tricky for, say, a surreptitious discovery of a nova or a new comet, though this hasn't happened (yet).

Astronomy is moving online in a big way, and online telescope services are an interesting option to keep eyes on the sky.

EPILOGUE

One more thing.

We've covered a good overview of amateur astronomy in this guide, from meteors, auroras, and satellites whizzing overhead just beyond the atmosphere, to quasars and galaxies in the most-distant corners of the cosmos and far back in space and time.

But when it comes to amateur astronomy and the study of the Universe, there's always something new, literally under (and beyond) the Sun. Sure, I like to revisit old friends such as Saturn and the Andromeda Galaxy whenever I can . . . but I also try to challenge myself to see at least one new object on every evening outing. This strategy has worked well for the half century I have shared this small corner of time and space, and I doubt I'll exhaust all the Universe has to offer in a single human lifetime.

My hope is that this guide will whet the appetite for something more. Astronomy and the pursuit of space and science are more than simple hobbies in search of novelty, such as stamp and bottle cap collecting. It also changes your outlook on life and your perspective on our place in time and space.

Here are just a few related topics we only touched on in the guide as they relate to amateur astronomy, all of which are topics which could be worthy of books in and of themselves:

- Mythology and the legends and lore of the sky
- Descriptive astrophysics
- A deep dive into the fascinating history of astronomy
- Calendars, timekeeping, and the cycles of astronomy and the sky
- Pronunciation for constellations, star names, and more
- Views in time and space, along with rare future and past events.

These are just a few avenues of interest to pursue, and maybe one day (in another book) we will.

Maybe you're just happy taking Milky Way landscape shots, or maybe just finding Saturn and showing it off to the public is your noble goal. Certainly, amateur astronomy has changed incredibly as a whole over the last few years, and will continue to do so.

Just a few short decades ago, a 6-inch (15-cm) reflector was a big telescope, film exposures for deep sky astrophotography were measured in hours, and information was often limited to the two dated books on astronomy found at the local library. Today, digital photography and the internet has changed up the game, though the goal always remains the same: getting to intimately know the night sky on your very own personal odyssey of discovery as a carbon-based life form, a child of the Universe getting to know itself.

Where is the game headed? Well, there are signs that electronic eyepieces and sensors may soon become the new game changers in amateur astronomy. Online, new surveys such as the LSST and the Transiting Exoplanet Survey Satellite (TESS) are set to release a torrent of data, information which will no doubt keep online sleuths up late at night looking for new exoplanets, asteroids, comets, and who knows what else. And the unpredictability is all part of the fun of the game. Sure, we know when eclipses and occultations will occur, but who could have predicted Hanny's Voorwerp, the hunt for exoplanets in Kepler data, or the explosion of an asteroid over Chelyabinsk one cold, clear February morning?

The drama of the Universe continues to unfold all around us every day, if we only know when and where to look for it.

BUT WHEN IT COMES TO AMATEUR ASTRONOMY

and the study of the Universe, there's always something new

CONSTELLATIONS OF THE SKY

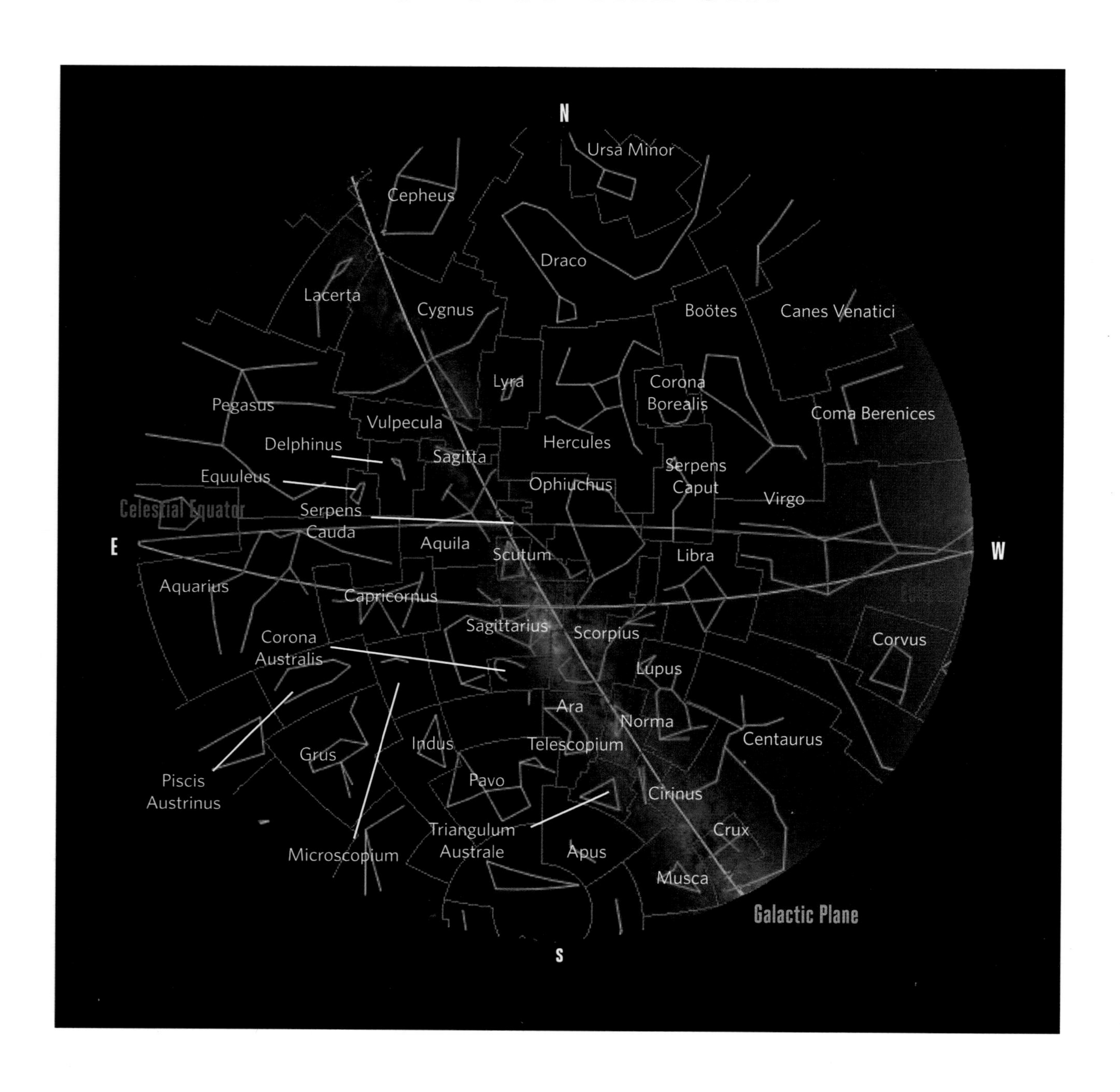

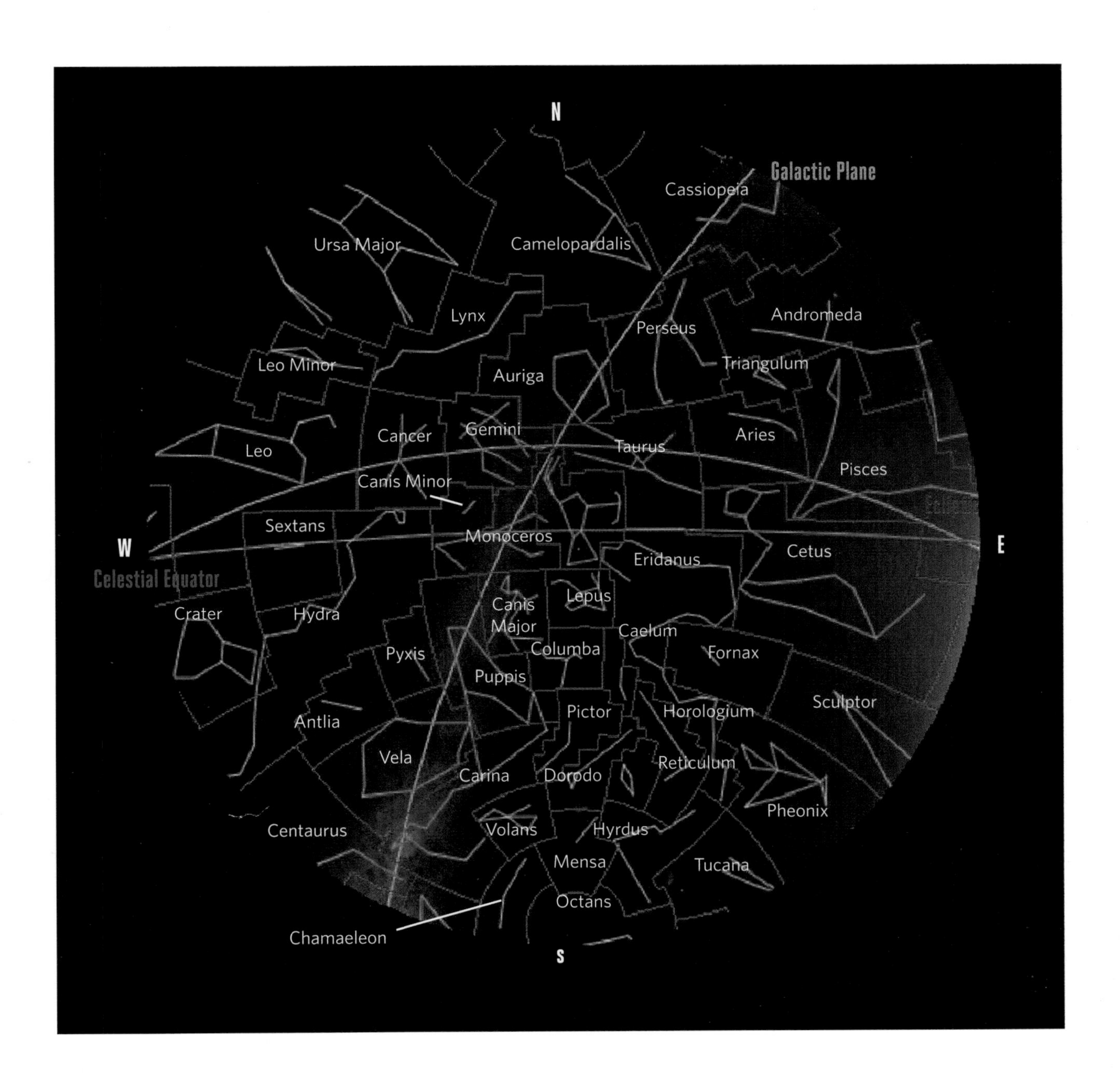
N
Galactic Plane
Cassiopeia
Ursa Major
Camelopardalis
Lynx
Andromeda
Perseus
Leo Minor
Triangulum
Auriga
Gemini
Aries
Cancer
Leo
Taurus
Pisces
Canis Minor
Sextans
Monoceros
W
E
Cetus
Eridanus
Celestial Equator
Crater
Hydra
Canis Major
Lepus
Caelum
Columba
Fornax
Pyxis
Puppis
Pictor
Horologium
Sculptor
Antlia
Vela
Reticulum
Carina
Dorodo
Pheonix
Centaurus
Volans
Hyrdus
Mensa
Tucana
Octans
Chamaeleon
S

ENDNOTES

CHAPTER 1

1. No Author Given, "Rods & Cones," *Rochester Institute for Technology Center for Imaging Science*. Retrieved November 25, 2017. https://www.cis.rit.edu/people/faculty/montag/vandplite/pages/chap_9/ch9p1.html

2. Eileen Hawley et al, "Expedition 1 Press Kit," *NASA*. October 25, 2000. https://spaceflight.nasa.gov/station/crew/exp1/exp1_presskit.pdf

3. No Author Given, "Distance to the Horizon Calculator," *Ringbell UK*. Retrieved December 1, 2017. http://www.ringbell.co.uk/info/hdist.htm

4. No Author Given, "A Brief History of Space," *Institute of Physics*. Retrieved January 1, 2018. http://www.iop.org/resources/topic/archive/space/

5. Fraser Cain, "How Far is Mars from the Earth?" *Universe Today*. December 5, 2013. https://www.universetoday.com/14824/distance-from-earth-to-mars/

6. Chris Peat, "Spacecraft Escaping the Solar System," *Heavens-Above*. Retrieved December 15, 2017. http://www.heavens-above.com

7. John E. Bortle, "The Bortle Sky Scale," *Sky & Telescope*. July 18, 2006. http://www.skyandtelescope.com/astronomy-resources/light-pollution-and-astronomy-the-bortle-dark-sky-scale/

8. Constellation Guide, "Southern Constellations," *Constellations: A Guide to the Night Sky*. Retrieved December 31, 2017. http://www.constellation-guide.com/constellation-map/southern-constellations/

9. Dava Sobel, *Longitude: The True Story of the Lone Genius Who Solved the Greatest Scientific Problem of His Time* (New York: Bloomsbury Press, 2010), 21.

10. No Author Given, "The Distance to the Center of the Milky Way Galaxy," *Harvard Center for Astrophysics*. Retrieved January 2, 2018. https://www.cfa.harvard.edu/~reid/trigpar.html

11. Brooke Boen, "Supermassive Black Hole Sagittarius A*," *NASA*. August 7, 2017. https://www.nasa.gov/mission_pages/chandra/multimedia/black-hole-SagittariusA.html

12. Astronomers have a skewed perspective on the periodic table: there's hydrogen, helium (non-metals), then the small fraction of the remainder of the Universe composed of heavier elements (metals).

13. Elizabeth Howell, "Vega, the Once and Future Pole Star," *Space.com*. June 25, 2013. https://www.space.com/21719-vega.html

14. Bruce McClure, "Apex of the Sun's Way," *I Dial Stars*. July 1, 2003. http://www.idialstars.com/apex.htm

15. Sol Station, "Barnard's Star." Retrieved December 20, 2017. http://www.solstation.com/stars/barnards.htm

16. Alan MacRobert, "The Stellar Magnitude System," *Sky & Telescope*. August 1, 2006. http://www.skyandtelescope.com/astronomy-resources/the-stellar-magnitude-system/

17. The International Comet Quarterly, "The Magnitude Scale." Retrieved December 28, 2017. http://www.icq.eps.harvard.edu/MagScale.html

18. Las Cumbres Observatory, "What is Absolute Magnitude?" Retrieved December 26, 2017. https://lco.global/spacebook/what-absolute-magnitude/

19. ASCOM: "Standards for Astronomy." Retrieved December 21, 2017. http://www.ascom-standards.org/

CHAPTER 2

1. Andrew Fazekas, "World's Largest Backyard Telescope," *National Geographic Magazine*. November 8, 2013. https://voices.nationalgeographic.org/2013/11/08/worlds-largest-backyard-telescope/

2. Roger Sinnott, "Where Did the 1¼-inch Standard Size for Eyepiece Barrels Come From?" *Sky and Telescope*. July 21, 2006. http://www.skyandtelescope.com/astronomy-resources/astronomy-questions-answers/where-did-the-1-14-inch-standard-size-for-eyepiece-barrels-come-from/

3. John Savard, "Eyepieces." Retrieved October 20, 2017. http://www.quadibloc.com/science/opt04.htm

4. Damian Peach, "Understanding Resolution and Contrast," *Views of the Solar System*. January 1, 2004. http://www.damianpeach.com/simulation.htm

5. Mansurov Nasim. "What is Chromatic Aberration?" *Photography Life*. March 2, 2017. https://photographylife.com/what-is-chromatic-aberration

6. No Author Given, "A Brief History of Telescopes," *The Sandy Historical Society*. May 13, 2014. http://sandyhistorical.org/2014/05/13/a-brief-history-of-optical-telescopes/

7. No Author Given, "The 40-Inch Telescope from the Inside," *University of Chicago*. April 20, 1999. http://astro.uchicago.edu/vtour/40inch/

8. No Author Given, "Newton's Reflectors," *Amazing Space*. Retrieved October 20, 2017. http://amazingspace.org/resources/explorations/groundup/lesson/scopes/newton/index.php

9. Ken Diller, "An Introduction to Newtonian Telescope Design," *Brevard Astronomical Society*. September 21, 2011. http://www.brevardastro.org/documents/Intro percent20to percent20Newt percent20Tel percent20Design.pdf

10. Geoff Barker, "Speculum Mirror Made for a Reflecting Telescope by Sir William Herschel," *Museum for Applied Arts and Sciences*. Retrieved October 19, 2017. https://collection.maas.museum/object/231971

11. Joseph Castro, "Who Invented the Mirror?" *Live Science*. March 28, 2013. https://www.livescience.com/34466-who-invented-mirror.html

12. Gary Kronk, "17/P Holmes," *Cometography*. Retrieved October 19, 2017. http://cometography.com/pcomets/017p.html

13. Alan Hale, "The Discovery of Comet Hale-Bopp," *NASA-JPL*. September 1, 1995. https://www2.jpl.nasa.gov/comet/discovery.html

14. Emily Lakdawalla, "The Jupiter Impact," *The Planetary Society*. August 13, 2009. http://www.planetary.org/blogs/emily-lakdawalla/2009/2049.html

15. Relive this "one that got away" live (right around 22:00) at: https://youtu.be/ePOrciuULH0

CHAPTER 3

1. No Author Given, "Planets and the Celestial Sphere," *University of Florida*. Retrieved October 30, 2017. http://www.astro.ufl.edu/~guzman/ast1002/class_notes/Ch1/Ch1.html

2. W.T. Lynn, "Copernicus and Mercury," *The Observatory*. August 1892. http://adsabs.harvard.edu/full/1892Obs....15..321L

3. Fred Espenak, "Transits of Mercury and Venus," *NASA/GSFC*. February 24, 2012. https://eclipse.gsfc.nasa.gov/transit/transit.html

4. Tony Phillips, "Venus Meets a Planet Named George," *NASA Science Beta*. April 11, 2006. https://science.nasa.gov/science-news/science-at-nasa/2006/11apr_george

5. IAU General Assembly, press release. "Result of IAU Resolution Votes." August 24, 2006. https://www.iau.org/news/pressreleases/detail/iau0603/

6. Thierry Legault, "Eclipsing the Sun," *Astronomy Picture of the Day*. January 5, 2011. https://apod.nasa.gov/apod/ap110105.html

7. C. Seligman, "Retrograde Motion," *Online Astronomy, Orbital Motions*. Retrieved November 1, 2017. https://cseligman.com/text/sky/retrograde.htm

8. Paolo Palmieri, "Galileo and the Discovery of the Phases of Venus," *Journal for the History of Astronomy*. 2001. http://adsabs.harvard.edu/full/2001JHA . . . .32..109P

9. The Sun is also increasing slightly in size and luminosity, accelerating this process as well.

10. Fred Schaaf, *The Brightest Stars* (Hoboken, New Jersey: John Wiley and Sons, Inc., 2008), 228.

11. Deborah Byrd, "Ancient Greek Coin May Mark the Blotting Out of Jupiter by the Moon," *Earth and Sky*. http://earthsky.org/human-world/ancient-greek-coin-might-mark-blotting-out-of-jupiter-by-the-moon

12. Kelly Beatty, "Global Fail for Big Regulus Cover Up," *Sky & Telescope*. March 27, 2014. http://www.skyandtelescope.com/astronomy-news/observing-news/global-fail-for-the-big-regulus-cover-up/

CHAPTER 4

1. The biggest change you'd notice in the sky if you time-travelled back a few thousand years is a shift in the poles due to the precession of the equinoxes.

2. The IAU also included eighty-eight more proper star names for good measure at the end of 2017: https://www.iau.org/news/pressreleases/detail/iau1707/

3. I like to refer to our year of surviving high school Latin as "the most informative class I ever struggled to get a C in . . ."

4. Harmurt Frommert, "History of the Discovery of Deep Sky Objects," *SEDS*. Retrieved November 17, 2017. http://www.messier.seds.org/xtra/history/deepskyd.html

5. Hunters of deceased celebrities at the Père Lachaise Cemetery in Paris, France, headed to Oscar Wilde's or Jim Morrison's grave may also want to keep an eye out for Charles Messier's tomb nearby.

6. Tammy Plotner, "Messier 40: the Winnecke 4 Double Star," *Universe Today*. May 8, 2017. https://www.universetoday.com/34324/messier-40-1/

7. Jacoby Suzanne, "AURA's Large Synoptic Survey Telescope to Begin Construction," *A&M Science*. August 4, 2014. http://www.science.tamu.edu/news/story.php?story_ID=1255#.Wg4hxUpKtRY

8. Larry McHenry, "Robert Trumpler," *Stellar Journeys*. Retrieved November 11, 2017. https://www.stellar-journeys.org/cluster-trumpler-info.htm

9. Lacaille was either unimaginative or simply enamored with eighteenth-century technology, forever enshrining such curious but unpoetic objects as Antila (the Pump) and Fornax (the Furnace) in the southern sky.

10. Elizabeth Howell, "Globular Clusters: Dense Groups of Stars," *Space.com*. July 22, 2015. https://www.space.com/29717-globular-clusters.html

11. As a child of the 1970s, I remember seeing astronomy textbooks stating the age of the Universe as twenty billion years, "plus or minus 50 percent."

12. As mentioned previously, the bright stars Spica and Betelgeuse are prime nearby supernova candidates.

CHAPTER 5

1. Many early observatories, such as the Leviathan of Parsonstown, were simple transit instruments, capable of only tracking objects as they briefly crossed the local meridian.

2. "Zero hour" for right ascension currently falls in the astronomical constellation of Pisces until around 2590 AD, when the northward equinoctial point moves into Aquarius. I wouldn't hold your breath for peace and understanding, though.

3. Deborah Byrd, "Which Arm of Our Milky Way Galaxy Contains the Sun?" *Earth and Sky*. May 20, 2014. http://earthsky.org/space/does-our-sun-reside-in-a-spiral-arm-of-the-milky-way-galaxy

4. The Jesuits actually have a long tradition of astronomy, and the Catholic Church still maintains professional observatories outside of Rome and on Mount Graham in Southern Arizona.
5. Felicia Chou, "Hubble Goes High-Definition to Revisit the Pillars of Creation," *NASA*. January 5, 2015. https://www.nasa.gov/content/goddard/hubble-goes-high-definition-to-revisit-iconic-pillars-of-creation
6. Ben Sullivan, "The Controversy over the 'Wow!' Signal is Astronomy's Greatest Beef," *Motherboard*. June 12, 2017. https://motherboard.vice.com/en_us/article/wjqzmy/the-controversy-over-the-alien-wow-signal-is-astronomys-greatest-beef
7. Ethan Siegel, "A Four Star Controversy Resolved," *Starts With a Bang*. October 21, 2013. http://scienceblogs.com/startswithabang/2013/10/21/messier-monday-a-four-star-controversy-resolved-m73/
8. Next time you see a Subaru car, look at the logo. It's actually the Pleiades star cluster!
9. For the intents and purposes of this guide, we're listing the month when the particular object transits around 10 p.m. local time.
10. Coordinates on these tables are rounded to the nearest arcminute.
11. We say practically, as the local horizon does slightly curve away from the observer depending on elevation, and atmospheric refraction does come into play very near to the horizon. For now, we'll average that odd 50.0001 percent down to a tidy 50 percent.

CHAPTER 6

1. E. Bell II, "Sputnik I," *NASA Space Science Data Coordinated Archive*. Retrieved November 27, 2017. https://nssdc.gsfc.nasa.gov/nmc/spacecraftDisplay.do?id=1957-001B
2. Fun fact: Most folks who claimed they saw Sputnik on U.S. passes actually caught sight of the much-larger rocket booster that placed it in orbit.
3. No Author Given, "About Space Debris," *The European Space Agency*. April 17, 2017. http://www.esa.int/Our_Activities/Operations/Space_Debris/About_space_debris
4. W. Patrick McCray, *Keep Watching the Skies! The Story of Operation Moonwatch and the Dawn of the Space Age* (Princeton, New Jersey: Princeton University Press, 2008), 223.
5. Ella Davies, "The Place Furthest (sic) from Land is Point Nemo," *BBC Earth*. October 5, 2016. http://www.bbc.com/earth/story/20161004-the-place-furthest-from-land-is-known-as-point-nemo
6. Richard Baum, *The Haunted Observatory: Curiosities from the Astronomer's Cabinet* (Amherst, New York: Prometheus Books, 2007), 245.
7. This boost is fastest at the equator, at about 1,037 mph (1,669 kph). This is also why nations try to place launch sites as far south as they can, such as Cape Canaveral in Florida and the Baikonur Cosmodrome in Kazakhstan.
8. While the U.S. won't publish tracking TLEs for DoD and allied, friendly state military missions, it will do so for military launches of non-aligned states (e.g., Russia, China, North Korea, etc.).
9. No Author Given, "Types of Orbits," *NASA.gov*. Retrieved November 26, 2017. https://earthobservatory.nasa.gov/Features/OrbitsCatalog/page2.php

10. With stable orbits lasting millions of years, many of the satellites in geostationary orbit may also become the longest lasting testaments of our civilization. Two projects, the LAGEOS satellite launched in 1976 and the Last Pictures project aboard Echostar XVI contain message-in-a-bottle style mementos for any future space-junk salvagers.

11. Unfortunately, the new constellation of IridiumNEXT satellites won't flare in the same fashion.

12. According to Iridium CEO Matt Desch: "We have approval for up to 10 (original) satellites with lower fuel levels to take up to 25 years for reentry—so last (Iridium flare) in 2043?" Desch notes, however, that most of the remaining first generation Iridiums will probably begin tumbling after 2019.

13. John Pike, "Yaogan Ocean Surveillance Satellites," *Global Security.org.* December 28, 2014. https://www.globalsecurity.org/space/world/china/yaogan-noss.htm

14. Trudy Bell, Tony Phillips, "A Super Solar Flare," *NASA Science Beta.* May 6, 2008. https://science.nasa.gov/science-news/science-at-nasa/2008/06may_carringtonflare

15. Andrew Fazekas, "Auroras Make Weird Noises, and Now We Know Why," *National Geographic.* June 2016. https://news.nationalgeographic.com/2016/06/auroras-sounds-noises-explained-earth-space-astronomy/

CHAPTER 7

1. David Eicher, *Comets! Visitors from Deep Space* (Cambridge, United Kingdom: Cambridge University Press, 2013), 42.

2. Halley's Comet also sparked the first modern-day comet hysteria as well. Astronomers had just discovered cyanogen—a deadly gas—in the tenuous tail of the comet, and the Earth passed through the vacuous tail of the comet on the night of May 19, 1910.

3. Hale-Bopp, for example, entered the Solar System on a 4,200-year orbit, and exited on a 2,533 year one.

4. Alas, with twenty-six letters in the alphabet but only 24 half-months, there will never again be a comet with a "Z" designation with the current system . . . "I" is skipped, lest it be confused with "1."

5. Brian Marsden, "Sungrazing Comets," *Annual Review of Astronomy and Astrophysics*. January 2015. http://www.annualreviews.org/doi/abs/10.1146/annurev.astro.43.072103.150554

6. This is even true after a comet rounds the Sun and heads back out of the Solar System, tail first!

7. CBAT actually received its last "telegram" in 1995, marking the discovery of Comet Hale-Bopp.

8. Christine Gorman, "The Comet that Battered Jupiter and Shook Congress," *Scientific American.* February 2016. https://www.scientificamerican.com/article/s19-the-comet-that-battered-jupiter-and-shook-congress/

9. The Perseids are also sometimes referred to as the Tears of Saint Lawrence, who was martyred on a hot gridiron on August 10, 258 AD.

10. James Richardson, "Fireball FAQs," *The American Meteor Society*. Retrieved December 7, 2017. https://www.amsmeteors.org/fireballs/faqf/

11. Sarah Zielinski, "The Rare Meteor Event that Inspired Walt Whitman," *Smithsonian Magazine*. June 2010. https://www.smithsonianmag.com/science-nature/rare-meteor-event-inspired-walt-whitman-29643165/

12. Sarah Lewin, "Company Aims to Offer On-Demand Meteor Showers," *Space.com*. July 2015. https://www.space.com/29849-on-demand-meteror-showers-ale.html

13. Colin Keay, "Meteor Fireball Sounds Identified," *SAO/NASA Astrophysics Data System*. 1992. http://adsabs.harvard.edu/full/1992acm..proc..297K

14. No Author Given, "How to Use Your FM Radio to Detect Meteor Showers," *Sky Scan Awareness Project.* November 2011. http://www.skyscan.ca/meteor_radio_detection.htm

15. About 1 percent of the background static on the FM dial is due to the Cosmic Background Radiation left over from the Big Bang, originating 13.7 billion years ago.

CHAPTER 8

1. Nola Taylor Redd, "Giant Stars: Facts, Future and the Fate of the Sun," *Space.com*. August 21, 2013. https://www.space.com/22471-red-giant-stars.html

2. Photons of light from the Sun make quite a journey, taking over 10,000 years (!) to escape the dense core, then only 8 minutes to traverse the gulf of space from the Sun to the Earth. The light shining on us is ancient, indeed.

3. Even heavy radioactive elements trace their origin back to supernovas and massive neutron star mergers.

4. Jennifer Bergman, "History of Sunspot Observations," *Windows to the Universe*. September 6, 2005. https://www.windows2universe.org/sun/activity/sunspot_history.html

5. Al Van Helden, "Thomas Harriot: 1560–1621," *The Galileo Project*. January 1995. http://galileo.rice.edu/sci/harriot.html

6. The oft repeated tale that Galileo went blind observing the Sun is apocryphal, as the famous astronomer developed glaucoma and cataracts in old age. One true victim of permanent "sun blindness" was the 19th century psychologist Gustav Fechner, who was fascinated by the afterimages he saw after staring at the Sun.

7. Delores Knipp, "Ms. Hisako Koyama: From Amateur Astronomer to Long-Term Solar Observer," *Space Weather AGU Journal*. October 2, 2017. http://onlinelibrary.wiley.com/doi/10.1002/2017SW001704/full

8. Tony Phillips, "Severe Space Weather: Social and Economic Impacts," *NASA Science Beta*. January 21, 2009. https://science.nasa.gov/science-news/science-at-nasa/2009/21jan_severespaceweather

9. Alex Young, "Solar Eclipse Safety," *NASA Solar Eclipse 2017*. Retrieved December 14, 2017. https://eclipse2017.nasa.gov/safety

10. No Author Given, "History's Biggest Sunspots," *SpaceWeather.com*. Retrieved December 13, 2017. http://spaceweather.com/sunspots/history.html

CHAPTER 9

1. The 1840 First Quarter Moon image was taken from a rooftop observatory on the New York University campus.

2. Monica Young, "Tabby's Star: Weird Star Gets Weirder," *Sky & Telescope*. August 12, 2016. http://www.skyandtelescope.com/astronomy-news/tabbys-star-weird-star-gets-weirder/

3. Alan Hirshfeld, *Starlight Detectives: How Astronomers, Inventors, and Eccentrics Discovered the Modern Universe* (New York: Bellevue Literary Press, 2014), 115.

4. Cory Schmitz, "Single vs. Multiple Stacking Exposures: is Stacking Worth It?" *Photographing Space*. Retrieved December 28, 2017. https://photographingspace.com/stacking-vs-single/

5. Mamta Patel Nagaraja, "Space Station Astrophotography," *Science@NASA*. March 24, 2003. https://science.nasa.gov/science-news/science-at-nasa/2003/24mar_noseprints

6. The northern celestial pole is actually currently 40 arcminutes from the North Star Polaris, in the direction of the star Kocab, which forms the top corner of the bowl of the Little Dipper asterism in the extended Ursa Minor constellation.

7. Martin Lewis, "How to Make a Simple All-Sky Camera," *Sky at Night Magazine*. February 22, 2016. http://www.skyatnightmagazine.com/feature/how-guide/how-make-simple-all-sky-camera

8. The maximum length of totality for the April 8, 2024, total solar eclipse is 4 minutes, 28 seconds.

9. This is because although the 99 percent illuminated Full Moon is nearly double the 50 percent illuminated extent of the First Quarter Moon, sunlight also strikes the Full Moon almost straight on from the zenith (from the perspective of the surface of the Moon), and features on the Moon around this time are almost devoid of shadows.

10. J. Kanipe and T. Webb, *Annals of the Deep Sky* (Richmond, Virginia: Willmann-Bell, Inc., 2015), vol. 1, 70.

11. Steve Cullen, "Blurred Lines Between Science and Art in Photography," *Fstoppers*. December 21, 2017. https://fstoppers.com/originals/blurred-lines-between-science-and-art-photography-208819

The stages of a lunar eclipse, from partial to totality.

CHAPTER 10

1. John Wenz, "Two Stars Will Merge in 2022 and Explode in Red Fury," *Astronomy Magazine* online. January 6, 2017. http://www.astronomy.com/news/2017/01/2022-red-nova

CHAPTER 11

1. Deborah Byrd, "Chelyabinsk Meteor Mystery 3 Years Later," *Earth Sky*. February 15, 2016. http://earthsky.org/space/chelyabinsk-meteor-mystery-3-years-later
2. Peter Jenniskens, "The Established Meteor Showers as Observed by CAMS," *Icarus*. September 15, 2015. http://cams.seti.org/ICARUS-CAMSI.pdf
3. The prototypical Cepheid variable from which the class of stars derives its name is Delta Cephei in the constellation of Cepheus, the King.
4. Mike Simonsen, "Amateurs Discover U Scorpii Eruption," *Sky & Telescope*. January 29, 2010. http://www.skyandtelescope.com/astronomy-news/amateurs-discover-u-scorpiis-eruption/
5. Daniel Clery, "Colliding Stars Will Light up the Night Sky in 2022," *Science Mag.org*. January 6, 2017. http://www.sciencemag.org/news/2017/01/colliding-stars-will-light-night-sky-2022
6. Ted Forte, "The Inconstancy of Nebulae," *Sky & Telescope*, 22 (February 2018).
7. Charles Choi, "Alien City Lights Could Signal E.T. Planets," *Space.com*. November 12, 2011. https://www.space.com/13514-alien-city-artificial-lights-extraterrestrial-planets.html
8. Sharon Guynup, "Light Pollution Taking a Toll on Nightlife, Eco-Groups Say," *National Geographic.com*. April 17, 2003. https://news.nationalgeographic.com/news/2003/04/0417_030417_tvlightpollution.html

The California Nebula: NGC 1499, an emission nebula located in the constellation Perseus, about 1,000 light-years from Earth. Imaged using narrowband filters: hydrogen is represented in yellow and oxygen in blue /white.

RESOURCES

CHAPTER 1 – STEPPING OUT TONIGHT

Clear Sky Chart weather forecast: http://www.cleardarksky.com/

The International Dark Sky Association (IDA): http://www.darksky.org/idsp/

Skippy Sky astronomical weather forecast: http://www.skippysky.com.au/

Stellarium freeware planetarium program: http://stellarium.org/

World Atlas of Artificial Sky Brightness: http://www.inquinamentoluminoso.it/worldatlas/pages/fig1.htm

CHAPTER 2 – ASTRONOMY GEAR AND TECH TALK

The Washington Double Star Catalog: an exhaustive list of double stars maintained by the United States Naval Observatory. http://ad.usno.navy.mil/wds/

CHAPTER 3 – FOLLOWING PLANETS AND THE MOON IN THE SKY

The Association of Lunar & Planetary Observers (ALPO): http://alpo-astronomy.org/

Eclipse Maps (maintained by Michael Zeiler): http://eclipse-maps.com/Eclipse-Maps/Welcome.html

Eclipse Weather (maps and weather predictions for future solar eclipses): http://eclipsophile.com/

NASA/Goddard Space Flight Center's Eclipse Web page: https://eclipse.gsfc.nasa.gov/eclipse.html

The USGS AstroGeology Site: https://astrogeology.usgs.gov/maps

CHAPTER 4 – THE DEEP SKY

Dark nebulae: http://www.skyandtelescope.com/observing/celestial-objects-to-watch/seeking-summers-dark-nebulae/

Interactive NGC catalog: http://spider.seds.org/ngc/ngc.html

Messier's catalog online: http://www.messier.seds.org/

CHAPTER 5 – THE SKY FROM SEASON TO SEASON

The American Association of Variable Star Observers: Complete with alerts and finder charts for new novae and supernovae. www.aavso.org

The best variables in the sky: http://www.astronomy.com/observing/get-to-know-the-night-sky/2006/12/fun-with-double-and-variable-stars

CHAPTER 6 – NEAR-SKY WONDERS

AURORA-OBSERVING RESOURCES

NASA: NASA's SOHO and SDO spacecraft monitor the Sun across the electromagnetic spectrum, around the clock. https://sohowww.nascom.nasa.gov/

https://sdo.gsfc.nasa.gov/

NOAA: A good place to check out the current state of the auroral oval, the Bz, and the Kp index. http://www.swpc.noaa.gov/

Space weather live: https://www.spaceweatherlive.com/en/solar-activity

Space weather watch: A good one-stop shopping site to monitor the space weather situation. http://www.spaceweather.com/

There are lots of free aurora alert apps out there. One of our faves is NASA/GFSC's Space Weather SWX, which will show you the current situation for SOHO, SDO, GOES, and Bz/Kp data.

SATELLITE-TRACKING RESOURCES

Thanks to the World Wide Web, satellite hunters have a wealth of information at their fingertips. Unlike other fields of astronomy, the world of satellite tracking can change quickly, with new launches going up and old space junk coming down.

Aerospace Corp: A California non-profit analysis group for the U.S. government. This site also publishes free-access reentry data for the public. http://www.aerospace.org/cords/reentry-predictions/

Celestrak: maintained by T.S. Kelso, Celestrak is a great open source for satellite data. https://celestrak.com/

Heavens-Above: Strange to think we've been using Heavens-Above for tracking satellites for two decades now. Though Heavens-Above also has astronomical information on the Sun, Moon, planets, and more, its extensive satellite database is really what sets it apart. Chris Peat created and manages Heavens-Above. The website will easily show you the sky situation (including brighter satellites) for your time and location. The Heavens-Above app is also a handy utility in the field for immediate satellite identification. http://www.heavens-above.com/

Orbitron: Free to download. Orbitron creator Sebastian Stoff refers to the satellite tracking platform as cardware, meaning he appreciates a postcard from users worldwide. Orbitron is a powerful satellite tracking program, and you can use it in the field on a laptop, away from an internet connection. The key to accuracy is that the user must periodically update the TLE database (maintained in a separate .txt file), either automatically from Celestrak, or manually from Space-Track or Heavens-Above. http://www.stoff.pl/

Satflare: Another notable satellite tracking application. http://www.satflare.com/

SeeSat-L: The venerable message board with lots of discussions concerning the latest goings on in Earth orbit. http://www.satobs.org/seesat/seesatindex.html

SpaceFlightNow/Spaceflight101: Two of the best news resources online for spaceflight news and information, including launch schedules and news about missions. Spaceflight 101 is particularly useful if you're looking for in-depth information on satellite payloads, including, say, the name and purpose of *every* individual CubeSat on a 90 payload mission. http://spaceflightnow.com/

Space-Track: US JSpOC's clearinghouse for satellite information, including recent TLEs and reentries. Space-Track is not open source, but requires a login. The good news is the process of getting a login is rather painless and they welcome amateur satellite trackers, after assuring you're not an agent for a rogue state looking to use satellite data for nefarious purposes. Orbits and tracking TLEs evolve over time as new launches take to space, orbital burns raise and lower trajectories, and atmospheric drag and solar activity brings space junk back down. Space-Track is the source for the most recent TLEs that all other satellite tracking apps and platforms mine for information. When a change occurs, it percolates outward from Space-Track to all other users, meaning you're getting the most current information here. http://www.space-track.org/

SOCRATES: An interesting newcomer, SOCRATES (the Satellite Orbital Conjunction Reports Assessing Threatening Encounters in Space) tracks close upcoming debris conjunctions (passes) worldwide. http://celestrak.com/SOCRATES/

Spot the Station: NASA's no-fuss website for tracking the International Space Station worldwide. They also post tracking elements for cargo and crewed spacecraft chasing down or departing from the station as well. https://spotthestation.nasa.gov/

Twitter: A great social media source for fast-breaking information. @VirtualAstro, for example, curates and alerts followers to every visible pass of the International Space Station over the United Kingdom. I also tweet as @Astroguyz and curate all launches worldwide and cover high interest satellite passes for wherever in the world we currently happen to reside.

CHAPTER 7: COSMIC INTRUDERS: OBSERVING COMETS, ASTEROIDS, AND METEOR SHOWERS

RESOURCES AND STAYING ON TOP OF FAST-BREAKING NEWS

In May 1983, Comet C/1983 H1 IRAS-Araki-Alcock sped past the Earth, only 0.031 AU or 2.9 million miles (4.6 million km) distant. This was the fourth-closest recorded passage of a comet near Earth. The record holder is D/1770 L1 Lexell, which passed just 0.0151 AU away. I was in high school in the 1980s, and I remember that most of us heard of the swift passage several days *after* closest approach.

Fast-forward to 2017, and there are now a wealth of resources to track comets and assure that you're out there and ready for tomorrow night's (possible) bright comet:

The Central Bureau for Astronomical Telegrams: (via subscription) the point source for new comet and novae/supernovae discoveries. http://www.cbat.eps.harvard.edu/services/Subscriptions.html

Cometography: Another good resource on comets, maintained by Gary Kronk. http://cometography.com/

The Comet Observation database (COBS): This is the best place to quickly see what other observers are reporting in terms of comet brightness and activity. https://cobs.si/

The International Astronomical Union Minor Planet Center: lists new comet discoveries, soon after they're announced. https://www.minorplanetcenter.net/iau/lists/LastYear.html

NASA's JPL Horizons: Not only will this show you the orbital path of a new comet, but it will also allow you to generate ephemerides for new comets and asteroids in the database, listing coordinates in right ascension and declination by time for your location, handy for pointing your telescope at a fast-moving object. https://ssd.jpl.nasa.gov/?horizons

Weekly Information for Bright Comets: A good compendium of current and upcoming bright comets, complete with projected light curves. Maintained by the Japanese amateur astronomer Seiichi Yoshida. http://aerith.net/comet/weekly/current.html

CHAPTER 8 – SOLAR OBSERVING

Daily Mount Wilson observatory sunspot drawings site: http://obs.astro.ucla.edu/intro.html

NASA's Sun Funnel resource site: https://eclipse2017.nasa.gov/make-sun-funnel

CHAPTER 9 – ASTROPHOTOGRAPHY 101

AstroHutech: seller of modified, deep sky ready DSLRs. http://www.hutech.com/

Astronomy Picture of the Day (APOD): NASA's long-running website, frequently featuring amateur images. https://apod.nasa.gov/apod/astropix.html

Astrophotography Tips from *Sky & Telescope* magazine:

http://www.skyandtelescope.com/astronomy-resources/astrophotography-tips/

http://www.skyandtelescope.com/astronomy-resources/astrophotography-tips/deep-sky-with-your-dslr/

Astrophotography tutorials from Alan Dyer: http://www.amazingsky.com/free-tutorials.html

Deep Sky Stacker: http://deepskystacker.free.fr/english/index.htmlHarvard College

Eclipse Orchestrator: http://www.moonglowtechnologies.com/products/EclipseOrchestrator/index.shtml

K3CCD Tools: http://www.pk3.org/k3ccdtools/

NASA's Photographing Eclipses tutorial: https://eclipse.gsfc.nasa.gov/SEhelp/SEphoto.htmlObservatory

Astronomical Plate Stacks: http://tdc-www.harvard.edu/plates/

PEMPro, available for 30-day free trial from CCDWare: http://www.ccdware.com/products/pempro/features.cfm

PHDGuider autoguider resource: http://www.stark-labs.com/phdguiding.html

Photographer's Ephemeris: http://photoephemeris.com/

Photographing Space on getting diffraction spikes in images: https://photographingspace.com/howto-diffraction-spikes/

PhotoPills for Milky Way shooting: http://photographingspace.com/photopills-for-milkyway/

PIPP, a useful stacking program: https://sites.google.com/site/astropipp/

Polar Alignment tutorial: http://astropixels.com/main/polaralignment.html

Registax: a great freeware image stacking and processing program: https://www.astronomie.be/registax/

Space Weather: Also frequently features amateur pics. http://www.spaceweather.com/

The *Universe Today* Flickr pool: https://www.flickr.com/groups/universetoday/

The launch of STS-131 Space Shuttle Discovery at dawn, as seen from Hudson, Florida, about 100 miles to the west of the launch site.

CHAPTER 10 – TOP ASTRONOMY EVENTS FOR 2019–2024

The British Astronomical Association Comet Section's list of future comets: https://www.ast.cam.ac.uk/~jds/

Greatest Elongations for Mercury and Venus: http://www.jgiesen.de/skymap/MercuryVenus/

Heavens-Above's Table of Planetary oppositions: www.heavens-above.com

NASA/GFSC Eclipse Web page: https://eclipse.gsfc.nasa.gov/eclipse.html

Occult version 4.2.0 for future lunar occultations of bright stars and planets: http://www.lunar-occultations.com/iota/occult4.htm

CHAPTER 11 – REAL SCIENCE YOU CAN DO AND PROTECTING THE NIGHT SKY

The American Association for Variable Star Observer's manual for monitoring variable stars with a DSLR camera: https://www.aavso.org/dslr-observing-manual

Astronomer Pamela Gay on pro/am collaborations: http://www.skyandtelescope.com/get-involved/pro-am-collaboration/how-to-get-involved-in-amateur-research/

Cosmoquest for Citizen Science, including Moon Mappers, Image Detective, and more: https://cosmoquest.org/x/

iTelescope: http://www.itelescope.net/

Slooh: https://www.slooh.com/

TransitSearch.org: http://www.transitsearch.org/

The Zooniverse: https://www.zooniverse.org/

CONTRIBUTING PHOTOGRAPHERS

All photos by author or in the Public Domain except as noted below.

All star maps in the book were created by the author using Stellarium.

All brands are referred to in this book for review purposes only and are not endorsements.

Cover: The Trifid Nebula complex (Messier 20) in the constellation of Sagittarius. A fine example of a combination of emission, reflection, and dark nebulae, M20 lies about 9,000 light-years away. This composite luminance image combines images from the 8.2-meter Subaru Telescope and the Hubble Space Telescope. Color data by Martin Pugh, image assembly and processing by Robert Gendler.

INTRODUCTION

Page 2: Ollie Taylor/UK Night Photography. Instagram: @ollietaylorphotography

Page 5: Nathan Duncan. Instagram: @archioptic

Page 8: Mike Ince. Instagram: @spike.photos

CHAPTER 1

Page 12 (lower left): Carlos Guana (TheMalibuArtist.com) Instagram: @themalibuartist

Page 19: Monika Deviat. Instagram: @deviantoptiks

Page 23: Precession of the equinoxes. Left diagram created by the author using Stellarium; right diagram: NASA/GSFC/ Robert Simmon.

Page 26: Gianni Krattli. Instagram: @gianni_krattli_photography

CHAPTER 2

Page 28 (lower left): Grant Kasprowicz. Instagram: @grantkaspo

Page 28 (upper right), page 39: Ann Aldrich. Instagram: @jackrabbithomesteadcabins

Page 31 (center image): Chris Cook Photography (German Equatorial, www.cookphoto.com, Instagram: @cookphoto)

Page 33: Matthew Harbinson. Instagram: @cosmicwreckingball

Page 41 (left): Tony Mullen. Instagram: @tonysize

Page 41 (right): Sébastien Riethauser. Instagram: @_srphotos_

Page 43: Dylan O'Donnell. deography.com

CHAPTER 3

Page 46 (lower left), page 51: Roger Hutchison FRAS. Instagram: @rogerhutch931

Page 48: Paul Stewart. Instagram: @upsidedownastronomer

Page 48: Hershel's telescope. Credits: mirror photo: David Dickinson. Telescope sketch: Herschel, W., Bunce, J., & Walker, J. William Herschel's Twenty-Foot Reflecting Telescope (1794). (Public domain.)

Page 49: New Horizons/Pluto/NASA. Image credit: NASA/ JHUAPL/SwRI/public domain.

Page 53: The Apollo craters with finder. Large Moon image by the author, inset via NASA's Lunar Reconnaissance Orbiter. (Public domain.)

Page 54: Lunar Map with highlights and labels. Lunar disk via NASA (public domain), labels by the author.

Page 55: 3x Astronomy Sketches. Steve Brown. Instagram: @ sjb_astro

Page 57: Robert Krawczyk. Instagram: @robert_krawczyk

Page 58 (left): Damian Peach. www.damianpeach.com

Page 58 (right): Jan Veleba. Instagram: @veleba_astrophoto

Page 60: Black spot image on Jupiter. Credit: R. Evans/J. Trauger/H. Hammel/HST Comet Science Team/NASA.

Page 62, page 46 (lower right): Marko Korosec. Instagram: @markokorosecnet

Page 63: Zheng Zhi. Instagram: @lifelens

Page 66: Alex Conu. Instagram: @alexconu

Page 67: Typical occultation with explainer. Created by the author using Occult 4.3 software.

CHAPTER 4

Page 70 (lower left): Dustin Gibson/OPT Telescopes. Instagram: @gibsonpics

Page 70 (lower right), page 73: Tim Morrill. Instagram: @Timm1138.

Page 70 (upper right), page 76: Amit Kamble. Instagram: @amitkamblephotography.

Page 72: Credit: Painting circa 1770 by Nicolas Ansiaume (1729–1786). Image in the public domain.

Page 77: Sun's position in the galaxy. Credit: NASA/JPL-CALtech/ESO/R. Hurt.

Page 79: Messier marathon course of the night. Credit: Path created by the author, sky map graphics by NASA's Scientific Visualization Studio/Ernie Wright.

CHAPTER 5

Page 80 (lower right): Dan Sullivan. Instagram: @dansullivan

Page 80 (lower left), page 90: Dan Sullivan. Instagram: @dansullivan.

Page 80 (upper right), page 100: Peter Pat. Instagram: @astroscapepete.

Page 103: Dylan O'Donnell. deography.com

CHAPTER 6

Page 104 (upper right), page 110: Glenn Davis. Instagram: @Glenndavisphotos.

Page 104 (lower right), page 120: Göran Strand. Instagram: @astrofotografen.

Page 104: ISS image. Credit: STS-199/NASA.

Page 107: Point Nemo location. Credit: NASA/DSCVR.

Page 113: Dylan O'Donnell. Instagram: @dylan_odonnell_

Page 116: Heavens-Above satellite list + skychart. Credit: Heavens-Above

Page 117: Szabolcs Nagy. Instagram: @metrolinaszabi

Page 118: Timo Oksanen. Instagram: @timoksanen

Page 119: Carrington sketch. Credit: astronomer Richard Carrington, public domain.

Page 121: NOAA auroral oval. Credit: NOAA.

CHAPTER 7

Page 122 (right): Tom Masterson. Instagram: @transient_astro

Page 123: NASA Hayakutake image. Credit: NASA/Space Telescope Science Institute, image in the public domain.

Page 124: ESA Comet 67P image. Image credit: ESA/Rosetta/OSIRIS MPS Team.

Page 125: Graphic: JPL/NASA orbital trace of a comet through the inner solar system. Credit: NASA/Space Place.

Page 126: Comet W3 Lovejoy imaged from the ISS. Credit: Expedition 30/ISS/NASA.

Page 127: The track of a typical comet through the sky. Credit: NASA/Goddard Spaceflight Center/Alex Mellinger/Central Michigan University.

Page 128: Mike MacKinven. Instagram: @mack_photography_nz

Page 129: A typical SOHO sungrazer comet. Credit: NASA/ESA/SOHO.

Page 131: Tommy Eliassen. Instagram: @tommyeliassen

Page 133: Göran Strand. Instagram: @astrofotografen

Page 133: Meteor painting. Credit: Fredric Church, image in the public domain.

Page 135: Bryan M. Goff. Instagram: @bryangoffphoto.

CHAPTER 8

Page 140 (left): Göran Strand. Instagram: @astrofotografen

Page 140 (right), page 145: Paul Stewart. Instagram: @upsidedownastronomer

Page 140 (lower right), page 148: Rodrigo Pastor Pensa. Instagram: @vitopastorini

Page 142: NASA solar wind diagram. Image credit: NASA/Goddard Space Flight Center.

Page 142: Sunspot. Image credit: NASA's Solar Dynamics Observatory.

Page 143: Myscha Theriault. Instagram: @myschatheriault

Page 149: Dr. Pamela Gay/Cosmoquest. Instagram: @starstryder

CHAPTER 9

Page 150 (lower left), page 163: Matthew Newman. Instagram: @matthewnewmanphotography

Page 150 (lower right), page 155, page 166: Alan Dyer. Flickr: Amazing Sky Astrophotography. www.amazingsky.com

Page 150 (upper right), page 169: CorySchmitz PhotographingSpace.com. Instagram: @photographingspace

Page 152: First Image of Totality. Image credit: Johann Berkowski (Public domain.) (Thanks to Xavier Jubier for obtaining and curating this image.)

Page 153: Connor Matherne. Instagram: @AstroConnor

Page 157: Yuri Beletsky. Instagram: @yuribeletsky

Page 160: Jan Veleba. Instagram: @veleba_astrophoto

Page 161 (2x images), page 171: Andrew Symes. Instagram: @FailedProtostar

Page 168: Bray Falls. Instagram: @astrofalls

CHAPTER 10

Page 172 (lower right), page 197: Matt Beckmann. Instagram: @mattbeckmannimages

Page 172 (upper right), page 196: Boris Štromar. Instagram: @astrobobo.

Page 174 (lower image): Peter Rosén. Instagram: @pixmix_photo.

Page 176: Xavier Jubier. http://xjubier.free.fr.

Page 179: Shahrin Ahmad. Twitter: @shahgazer

Page 182: Everett Bloom. Instagram: @everett_bloom_photography

Page 184: ESA/Comet 67P. Image credit: ESA/Rosetta/NAVCAM.

Page 191: Michael Zeiler. Eclipse-Maps.com.

Page 192: Niels Haagh. Instagram: @astrodane

Page 196: Myscha Theriault. Instagram: @myschatheriault

CHAPTER 11

Page 198 (lower left): Marko Korosec. Instagram: @markokorosecnet

Page 198 (upper right), page 207: Cory Schmitz. www.photographingspace.com. Instagram: @photographingspace

Page 199: Göran Strand. astrofotografen.se

Page 203: Mars-moon eclipse MSL/NASA. (Public domain.) Image credit: NASA/JPL-Caltech/Texas A&M University/Malin Space Systems.

Page 204: Tanja Schmitz. www.astrotanja.com

Page 205: Light pollution image. Credit: NASA Earth Observatory/The Suomi NPP Satellite.

Page 212: Nathan Duncan. Instagram: @archioptic

Page 212: Ollie Taylor/UK Night Photography. Instagram: @ollietaylorphotography

ENDNOTES/CREDITS

Page 224: Dylan O'Donnell. Instagram: @dylan_odonnell_

Page 225: Stefano Cancelli. Instagram: @astro.garage

Page 234: NASA/HST/STScl. Processing: Judy Schmidt

ACKNOWLEDGMENTS

They say you have your entire life to write your first book, but you're on a deadline to write your second.

Lots of folks were either directly or indirectly involved with this project, too numerous to name. I'd like to thank Fraser Cain for giving me the final push needed to start this project and Page Street Publishing for believing in the vision. I'd like to thank Nancy Atkinson for giving me some insight into the first-time book-writing process, and Bob King for giving me some insight into how he makes his amazing graphics.

I'd also like to thank Christopher Becke and Cory Schmitz for the input and advice on the astrophotography chapter, one of the hardest and most technical chapters to put together. Also, thanks to everyone on social media who gave input on the samples shared.

Thanks to my mom, who instilled an interest in nature and astronomy in me at an early age, and who confiscated the dangerous screw-on eyepiece solar filter that came with my 60 mm Jason refractor, probably saving my eyesight in the process.

And finally, I'd like to thank my wife, Myscha, who tolerated my grunts, nods, and faint, unintelligible replies as we worked together through life's ups and downs as freelance nomads while I also worked on this book. Only a writer understands and recognizes the eccentric habits of another, and sees staring at a wall or out the window as actual, real *work* and all part of the process.

Thanks, Honey. I'm truly privileged to share this small corner of time and space with you.

ABOUT THE AUTHOR

Photo by Myscha Theriault. Instagram: @myschatheriault

David Dickinson is a longtime amateur astronomer of 40-plus years, a science writer, a science fiction author, and an enlisted U.S. Air Force veteran. He lives with his wife Myscha in Norfolk, Virginia. He is a frequent contributor to *Universe Today*, *Sky & Telescope*, and his own blog, www.astroguyz.com.

A mosaic of Hubble images of the Lagoon Nebula (Messier 8), an emission nebula 4,100 light-years away in the constellation Sagittarius.

INDEX

D

E

F

G

H